T0245187

The Skew-Normal and Related Families

Interest in the skew-normal and related families of distributions has grown enormously over recent years, as theory has advanced, challenges of data have grown and computational tools have become more readily available. This comprehensive treatment, blending theory and practice, will be the standard resource for statisticians and applied researchers. Assuming only basic knowledge of (non-measure-theoretic) probability and statistical inference, the book is accessible to the wide range of researchers who use statistical modelling techniques.

Guiding readers through the main concepts and results, the book covers both the probability and the statistics sides of the subject, in the univariate and multivariate settings. The theoretical development is complemented by numerous illustrations and applications to a range of fields including quantitative finance, medical statistics, environmental risk studies and industrial and business efficiency. The authors' freely available R package sn, available from CRAN, equips readers to put the methods into action with their own data.

ADELCHI AZZALINI was Professor of Statistics in the Department of Statistical Sciences at the University of Padua until his retirement in 2013. Over the last 15 years or so, much of his work has been dedicated to the research area of this book. He is regarded as the pioneer of this subject due to his 1985 paper on the skew-normal distribution; in addition, several of his subsequent papers, some of which have been written jointly with Antonella Capitanio, are considered to represent fundamental steps. He is the author or co-author of three books, over 70 research papers and four packages written in the R language.

ANTONELLA CAPITANIO is Associate Professor of Statistics in the Department of Statistical Sciences at the University of Bologna. She began working on the skew-normal distribution about 15 years ago, co-authoring with Adelchi Azzalini a series of papers, related to the skew-normal and skew-elliptical distributions, which have provided key results in this area.

INSTITUTE OF MATHEMATICAL STATISTICS
MONOGRAPHS

IMS Monographs are concise research monographs of high quality on any branch of statistics or probability of sufficient interest to warrant publication as books. Some concern relatively traditional topics in need of up-to-date assessment. Others are on emerging themes. In all cases the objective is to provide a balanced view of the field.

The Skew-Normal and Related Families

ADELCHI AZZALINI
Università degli Studi di Padova

with the collaboration of

ANTONELLA CAPITANIO
Università di Bologna

CAMBRIDGE
UNIVERSITY PRESS

University Printing House, Cambridge CB2 8BS, United Kingdom

One Liberty Plaza, 20th Floor, New York, NY 10006, USA

477 Williamstown Road, Port Melbourne, VIC 3207, Australia

314-321, 3rd Floor, Plot 3, Splendor Forum, Jasola District Centre, New Delhi - 110025, India

79 Anson Road, #06-04/06, Singapore 079906

Cambridge University Press is part of the University of Cambridge.

It furthers the University's mission by disseminating knowledge in the pursuit of
education, learning and research at the highest international levels of excellence.

www.cambridge.org
Information on this title: www.cambridge.org/9781108461139

First published 2014
First paperback edition 2018

A catalogue record for this publication is available from the British Library

Library of Congress Cataloging in Publication data
Azzalini, Adelchi, author.
The skew-normal and related families / Adelchi Azzalini, Università degli
Studi di Padova with the collaboration of Antonella Capitanio, Università di Bologna.
pages cm
Includes bibliographical references and index.
ISBN 978-1-107-02927-9 (Hardback)
1. Distribution (Probability theory) I. Capitanio, Antonella, 1964– author.
II. Title.
QA273.6.A98 2014
519.2´4–dc23   2013030070

ISBN 978-1-107-02927-9 Hardback
ISBN 978-1-108-46113-9 Paperback

Contents

Preface

Since about the turn of the millennium, the study of parametric families of probability distributions has received new, intense interest. The present work is an account of one approach which has generated a great deal of activity.

The distinctive feature of the construction to be discussed is to start from a symmetric density function and, by suitable modification of this, generate a set of non-symmetric distributions. The simplest effect of this process is represented by skewness in the distribution so obtained, and this explains why the prefix 'skew' recurs so often in this context. The focus of this construction is not, however, skewness as such, and we shall not discuss the quintessential nature of skewness and how to measure it. The target is instead to study flexible parametric families of continuous distributions for use in statistical work. A great deal of those in standard use are symmetric, when the sample space is unbounded. The aim here is to allow for possible departure from symmetry to produce more flexible and more realistic families of distributions.

The concentrated development of research in this area has attracted the interest of both scientists and practitioners, but often the variety of proposals and the existence of related but different formulations bewilders them, as we have been told by a number of colleagues in recent years. The main aim of this work is to provide a key to enter this theme. Besides its role as an introductory text for the newcomer, we hope that the present book will also serve as a reference work for the specialist.

This is not the first book covering this area: there exists a volume, edited by Marc Genton in 2004, which has been very beneficial to the dissemination of these ideas, but since its publication many important results have appeared and the state of the art is now quite different. Even today a definitive stage of development of this field has not been reached, if one assumes for a moment that such a state can ever be achieved, but we feel that the material is now sufficiently mature to also be fruitfully used for routine work of non-specialists.

The general framework and the key concepts of our development are formulated in Chapter 1. Subsequent chapters develop specific directions, in the univariate and in the multivariate case, and discuss why other directions are given lesser importance or even neglected. Some people may find it surprising that quite ample space is given to univariate distributions, considering that the context of multivariate distributions is where the new proposals appear more significant. However, besides its interest *per se*, the univariate case facilitates the exposition of many concepts, even when their main relevance is in the multivariate context.

There is a noticeable difference in the more articulate expository style of Chapters 1 to 6 compared with the briefer – even meagre one might say – summaries employed in Chapters 7 and 8, which deal with more specific themes. One reason for this choice is the greater importance given to the exposition of the basic concepts, recalling our main target in writing the book, and certain applied topics do not require a detailed discussion after the foundations of the construction are in place. Moreover, some of the more specialized or advanced topics are still in an evolutionary state, and any attempt to arrange them in an organized system is likely to become obsolete quite rapidly.

Chapters 1 to 6 are organized with a set of complements each, dealing with some more specialized topics. At first reading or if a reader is interested in getting a grasp of the key concepts only, these complements can be skipped without hindrance to understanding the core parts. At the end of these chapters there are sets of problems of varied levels of difficulty. As a rule of thumb, the harder ones are those with a reference at the end.

The development of this work has greatly benefited from the generous help of Giuliana Regoli, who has dedicated countless hours to examining and discussing with us many mathematical aspects. Obviously, any remaining errors are our own responsibility. We are also grateful to Elvezio Ronchetti, Marco Minozzo and Chris Adcock for comments on aspects of robustness, time series and quantitative finance, respectively, and to Marc Genton for several remarks on the nearly final draft. Even if in a less tangible form, our views on this research area have benefited from interactions with people of the 'skew community', with whom we have shared our enthusiasm during these years. It has been a stimulating and rewarding enterprise.

Adelchi Azzalini and Antonella Capitanio

February 2013

1

Modulation of symmetric densities

1.1 Motivation

This book deals with a formulation for the construction of continuous probability distributions and connected statistical aspects. Before we begin, a natural question arises: with so many families of probability distributions currently available, do we need any more?

There are three motivations for the development ahead. The first motivation lies in the essence of the mechanism itself, which starts with a continuous symmetric density function that is then modified to generate a variety of alternative forms. The set of densities so constructed includes the original symmetric one as an 'interior point'. Let us focus for a moment on the normal family, obviously a case of prominent importance. It is well known that the normal distribution is the limiting form of many non-normal parametric families, while in the construction to follow the normal distribution is the 'central' form of a set of alternatives; in the univariate case, these alternatives may slant equally towards the negative and the positive side. This situation is more in line with the common perception of the normal distribution as 'central' with respect to others, which represent 'departures from normality' rather than 'incomplete convergence to normality'.

The second motivation derives from the applicability of the mechanism to the multivariate context, where the range of tractable distributions is much reduced compared to the univariate case. Specifically, multivariate statistics for data in Euclidean space is still largely based on the normal distribution. Some alternatives exist, usually in the form of a superset, of which the most notable example is represented by the class of elliptical distributions. However, these retain a form of symmetry and this requirement may sometimes be too restrictive, especially when considering that symmetry must hold for all components.

The third motivation derives from the mathematical elegance and tractability of the construction, in two respects. First, the simplicity and generality of the construction is capable of encompassing a variety of interesting subcases without requiring particularly complex formulations. Second, the mathematical tractability of the newly generated distributions is, at least in some noteworthy cases, not much reduced compared to the original symmetric densities we started with. A related but separate aspect is that these modified families retain some properties of the parent symmetric distributions.

1.2 Modulation of symmetry

The rest of this chapter builds the general framework within which we shall develop specific directions in subsequent chapters. Consequently, the following pages adopt a somewhat more mathematical style than elsewhere in the book. Readers less interested in the mathematical aspects may wish to move on directly to Chapter 2. While this is feasible, it would be best to read at least to the end of the current section, as this provides the core concepts that will recur in subsequent chapters.

1.2.1 A fairly general construction

Many of the probability distributions to be examined in this book can be obtained as special instances of the scheme to be introduced below, which allows us to generate a whole set of distributions as a perturbed, or modulated, version of a symmetric probability density function f_0, which we shall call the *base density*. This base is *modulated*, or *perturbed*, by a factor which can be chosen quite freely because it must satisfy very simple conditions.

Since the notion of symmetric density plays an important role in our development, it is worth recalling that this idea has a simple and commonly accepted definition only in the univariate case: we say that the density f_0 is symmetric about a given point x_0 if $f_0(x - x_0) = f_0(x_0 - x)$ for all x, except possibly a negligible set; for theoretical work, we can take $x_0 = 0$ without loss of generality. In the d-dimensional case, the notion of symmetric density can instead be formulated in a variety of ways. In this book, we shall work with the condition of central symmetry: according to Serfling (2006), a random variable X is centrally symmetric about 0 if it is distributed as $-X$. In case X is a continuous variable with density function denoted $f_0(x)$, then central symmetry requires that $f_0(x) = f_0(-x)$ for all $x \in \mathbb{R}^d$, up to a negligible set.

Proposition 1.1 *Denote by f_0 a probability density function on $\mathbb{R}^d$, by $G_0(\cdot)$ a continuous distribution function on the real line, and by $w(\cdot)$ a real-valued function on $\mathbb{R}^d$, such that*

$$f_0(-x) = f_0(x), \quad w(-x) = -w(x), \quad G_0(-y) = 1 - G_0(y) \qquad (1.1)$$

for all $x \in \mathbb{R}^d$, $y \in \mathbb{R}$. Then

$$f(x) = 2 f_0(x) G_0\{w(x)\} \qquad (1.2)$$

is a density function on $\mathbb{R}^d$.

Technical proof Note that $g(x) = 2 [G_0\{w(x)\} - \frac{1}{2}] f_0(x)$ is an odd function and it is integrable because $|g(x)| \le f_0(x)$. Then

$$0 = \int_{\mathbb{R}^d} g(x) \, dx = \int_{\mathbb{R}^d} 2 f_0(x) G_0\{w(x)\} \, dx - 1 . \qquad \text{QED}$$

Although this proof is adequate, it does not explain the role of the various elements from a probability viewpoint. The next proof of the same statement is more instructive. In the proof below and later on, we denote by $-A$ the set formed by reversing the sign of all elements of A, if A denotes a subset of a Euclidean space. If $A = -A$, we say that A is a symmetric set.

Instructive proof Let Z_0 denote a random variable with density f_0 and T a variable with distribution G_0, independent of Z_0. To show that $W = w(Z_0)$ has distribution symmetric about 0, consider a Borel set A of the real line and write

$$\mathbb{P}\{W \in -A\} = \mathbb{P}\{-W \in A\} = \mathbb{P}\{w(-Z_0) \in A\} = \mathbb{P}\{w(Z_0) \in A\} ,$$

taking into account that Z_0 and $-Z_0$ have the same distribution. Since T is symmetric about 0, then so is $T - W$ and we conclude that

$$\tfrac{1}{2} = \mathbb{P}\{T \le W\} = \mathbb{E}_{Z_0}\{\mathbb{P}\{T \le w(Z_0)|Z_0 = x\}\} = \int_{\mathbb{R}^d} G_0\{w(x)\} f_0(x) \, dx .$$

$$\text{QED}$$

On setting $G(x) = G_0\{w(x)\}$ in (1.2), we can rewrite (1.2) as

$$f(x) = 2 f_0(x) G(x) \qquad (1.3)$$

where

$$G(x) \ge 0, \quad G(x) + G(-x) = 1 . \qquad (1.4)$$

Vice versa, any function G satisfying (1.4) can be written in the form $G_0\{w(x)\}$. For instance, we can set

$$
\begin{aligned}
G_0(y) &= \left(y + \tfrac{1}{2}\right) I_{(-1,1)}(2y) + I_{[1,+\infty)}(2y) \quad (y \in \mathbb{R}), \\
w(x) &= G(x) - \tfrac{1}{2} \qquad\qquad\qquad\qquad\qquad (x \in \mathbb{R}^d),
\end{aligned}
\tag{1.5}
$$

where $I_A(\cdot)$ denotes the indicator function of set A; more simply, this G_0 is the distribution function of a $\mathrm{U}(-\tfrac{1}{2}, \tfrac{1}{2})$ variate. We have therefore obtained the following conclusion.

Proposition 1.2 *For any given density f_0 in $\mathbb{R}^d$, such that $f_0(x) = f_0(-x)$, the set of densities of type (1.1)–(1.2) and those of type (1.3)–(1.4) coincide.*

Which of the two forms, (1.2) or (1.3), will be used depends on the context, and is partly a matter of taste. Representation of $G(x)$ in the form $G_0\{w(x)\}$ is not unique since, given any such representation,

$$
G(x) = G_*\{w_*(x)\}, \quad w_*(x) = G_*^{-1}[G_0\{w(x)\}]
$$

is another one, for any monotonically increasing distribution function G_* on the real line satisfying $G_*(-y) = 1 - G_*(y)$. Therefore, for mathematical work, the form (1.3)–(1.4) is usually preferable. In contrast, $G_0\{w(x)\}$ is more convenient from a constructive viewpoint, since it immediately ensures that conditions (1.4) are satisfied, and this is how a function G of this type is usually constructed. Therefore, we shall use either form, $G(x)$ or $G_0\{w(x)\}$, depending on convenience.

Since $w(x) = 0$ or equivalently $G(x) = \tfrac{1}{2}$ are admissible functions in (1.1) and (1.4), respectively, the set of modulated functions generated by f_0 includes f_0 itself. Another immediate fact is the following *reflection property*: if Z has distribution (1.2), $-Z$ has distribution of the same type with $w(x)$ replaced by $-w(x)$, or equivalently with $G(x)$ replaced by $G(-x)$ in (1.3).

The modulation factor $G_0\{w(x)\}$ in (1.2) can modify radically and in very diverse forms the base density. This fact is illustrated graphically by Figure 1.1, which displays the effect on the contour level curves of the base density f_0 taken equal to the $N_2(0, I_2)$ density when the perturbation factor is given by $G_0(y) = e^y/(1 + e^y)$, the standard logistic distribution function, evaluated at

$$
w(x) = \frac{\sin(p_1 x_1 + p_2 x_2)}{1 + \cos(q_1 x_1 + q_2 x_2)}, \qquad x = (x_1, x_2) \in \mathbb{R}^2, \tag{1.6}
$$

for some choices of the real parameters p_1, p_2, q_1, q_2.

Densities of type (1.2) or (1.3) are often called *skew-symmetric*, a term which may be surprising when one looks for instance at Figure 1.1, where

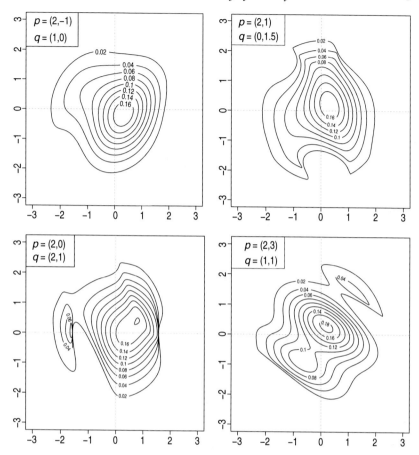

Figure 1.1 Density function of a bivariate standard normal variate with independent components modulated by a logistic distribution factor with argument regulated by (1.6) using parameters indicated in the top-left corner of each panel.

skewness is not the most distinctive feature of these non-normal distributions, apart from possibly the top-left plot. The motivation for the term 'skew-symmetric' originates from simpler forms of the function $w(x)$, which actually lead to densities where the most prominent feature is asymmetry. A setting where this happens is the one-dimensional case with linear form $w(x) = \alpha x$, for some constant α, a case which was examined extensively in the earlier stages of development of this theme, so that the prefix 'skew' came into use, and was later used also where skewness is not really the most distinctive feature. Some instances of the linear type will be discussed in detail later in this book, especially but not only in Chapter 2. However, in the more general context discussed in this chapter, the prefix 'skew' may be slightly misleading, and we prefer to use the term modulated or perturbed symmetry.

The aim of the rest of this chapter is to examine the general properties of the above-defined set of distributions and of some extensions which we shall describe later on. In subsequent chapters we shall focus on certain subclasses, obtained by adopting a specific formulation of the components f_0, G_0 and w of (1.2). We shall usually proceed by selecting a certain parametric set of functions for these three terms. We make this fact more explicit with notation of the form

$$f(x) = 2 f_0(x) G_0\{w(x; \alpha)\}, \qquad x \in \mathbb{R}^d, \qquad (1.7)$$

where $w(x; \alpha)$ is an odd function of x, for any fixed value of the parameter α. For instance, in (1.6) α is represented by (p_1, p_2, q_1, q_2). However, later on we shall work mostly with functions w which have a more regular behaviour, and correspondingly the densities in use will usually fluctuate less than those in Figure 1.1. In the subsequent chapters, we shall also introduce location and scale parameters, not required for the aims of the present chapter.

A word of caution on this programme of action is appropriate, even before we start to expand it. The densities displayed in Figure 1.1 provide a direct perception of the high flexibility that can be achieved with these constructions. And it would be very easy to proceed further, for instance by adding cubic terms in the arguments of $\sin(\cdot)$ and $\cos(\cdot)$ in (1.6). Clearly, this remark applies more generally to parametric families of type (1.7). However, when we use these distributions in statistical work, one must match flexibility with feasibility of the inferential process, in light of the problem at hand and of the available data. The results to be discussed make available powerful tools for constructing very general families of probability distributions, but power must be exerted with wisdom, as in other human activities.

1.2.2 Main properties

Proposition 1.3 (Stochastic representation) *Under the setting of Propositions 1.1 and 1.2, consider a d-dimensional variable Z_0 with density function $f_0(x)$ and, conditionally on Z_0, let*

$$S_{Z_0} = \begin{cases} +1 & \text{with probability } G(Z_0), \\ -1 & \text{with probability } G(-Z_0). \end{cases} \qquad (1.8)$$

Then both variables

$$Z' = (Z_0 | S_{Z_0} = 1), \tag{1.9}$$

$$Z = S_{Z_0} Z_0 \tag{1.10}$$

have probability density function (1.2). The variable S_{Z_0} can be represented in either of the forms

$$S_{Z_0} = \begin{cases} +1 & \text{if } T < w(Z_0), \\ -1 & \text{otherwise,} \end{cases} \qquad S_{Z_0} = \begin{cases} +1 & \text{if } U < G(Z_0), \\ -1 & \text{otherwise,} \end{cases} \tag{1.11}$$

where $T \sim G_0$ and $U \sim U(0,1)$ are independent of Z_0.

Proof First note that marginally $\mathbb{P}\{S = 1\} = \int_{\mathbb{R}^d} G(x) f_0(x)\, \mathrm{d}x = \frac{1}{2}$, and then apply Bayes' rule to compute the density of Z' as the conditional density of $(Z_0 | S = 1)$, that is

$$f_{Z'}(x) = \frac{\mathbb{P}\{S = 1 | Z_0 = x\}\ f_0(x)}{\mathbb{P}\{S = 1\}} = 2\, G(x)\, f_0(x)\,.$$

Similarly, the variable $Z'' = (Z_0 | S_{Z_0} = -1)$ has density $2\, G(-x)\, f_0(x)$. The density of Z is an equal-weight mixture of Z' and $-Z''$, namely

$$\tfrac{1}{2}\{2\, f_0(x)\, G(x)\} + \tfrac{1}{2}\{2\, f_0(-x)\, G(x)\} = 2\, f_0(x)\, G(x)\,.$$

Representations (1.11) are obvious. QED

An immediate corollary of representation (1.10) is the following property, which plays a key role in our construction.

Proposition 1.4 (Modulation invariance) *If the random variable Z_0 has density f_0 and Z has density f, where f_0 and f are as in Proposition 1.1, then the equality in distribution*

$$t(Z) \overset{\mathrm{d}}{=} t(Z_0) \tag{1.12}$$

holds for any q-valued function $t(x)$ such that $t(x) = t(-x) \in \mathbb{R}^q$, $q \geq 1$.

We shall refer to this property also as *perturbation invariance*. An example of the result is as follows: if the density function of the two-dimensional variable (Z_1, Z_2) is one of those depicted in Figure 1.1, we can say that $Z_1^2 + Z_2^2 \sim \chi_2^2$, since this fact is known to hold for their base density f_0, that is when $(Z_1, Z_2) \sim N_2(0, I_2)$ and $t(x) = x_1^2 + x_2^2$ is an even function of $x = (x_1, x_2)$.

An implication of Proposition 1.4 which we shall use repeatedly is that

$$|Z_r| \overset{\mathrm{d}}{=} |Z_{0,r}| \tag{1.13}$$

for the rth component of Z and Z_0, respectively, on taking $t(x) = |x_r|$. This fact in turn implies invariance of even-order moments, so that

$$\mathbb{E}\{Z_r^m\} = \mathbb{E}\{Z_{0,r}^m\}, \qquad m = 0, 2, 4, \ldots, \tag{1.14}$$

when they exist. Clearly, equality of even-order moments holds also for more general forms such as

$$\mathbb{E}\{Z_r^k Z_s^{m-k}\} = \mathbb{E}\{Z_{0,r}^k Z_{0,s}^{m-k}\}, \qquad m = 0, 2, 4, \ldots; \quad k = 0, 1, \ldots, m.$$

It is intuitive that the set of densities of type (1.2)–(1.3) is quite wide, given the weak requirements involved. This impression is also supported by the visual message of Figure 1.1. The next result confirms this perception in its extreme form: all densities belong to this class.

Proposition 1.5 *Let f be a density function with support $S \subseteq \mathbb{R}^d$. Then a representation of type (1.3) holds, with*

$$f_0(x) = \tfrac{1}{2}\{f(x) + f(-x)\},$$

$$G(x) = \begin{cases} \dfrac{f(x)}{2f_0(x)} & \text{if } x \in S_0, \\ \text{arbitrary} & \text{otherwise,} \end{cases} \tag{1.15}$$

where $S_0 = S \cup (-S)$ is the support of $f_0(x)$ and the arbitrary branch of G satisfies (1.4). Density f_0 is unique, and G is uniquely defined over S_0.

The meaning of the notation $-S$ is explained shortly after Proposition 1.1.

Proof For any $x \in S_0$, the identity

$$f(x) = 2\,\frac{f(x) + f(-x)}{2}\,\frac{f(x)}{f(x) + f(-x)}$$

holds, and its non-constant factors coincide with those stated in (1.15). To prove uniqueness of this factorization on S_0, assume that there exist f_0 and G such that $f(x) = 2 f_0(x) G(x)$ and they satisfy $f_0(x) = f_0(-x)$ and (1.4). From

$$f(x) + f(-x) = 2 f_0(x)\{G(x) + G(-x)\} = 2 f_0(x),$$

it follows that f_0 must satisfy the first equality in (1.15). Since $f_0 > 0$ and it is uniquely determined over S_0, then so is $G(x)$. QED

Rewriting the first expression in (1.15) as $f(-x) = 2 f_0(x) - f(x)$, followed by integration on $(-\infty, x_1] \times \cdots \times (-\infty, x_d]$, leads to

$$\overline{F}(-x) = 2 F_0(x) - F(x), \qquad x = (x_1, \ldots, x_d) \in \mathbb{R}^d, \tag{1.16}$$

if F_0 denotes the cumulative distribution function of f_0 and $\overline{F}$ denotes the survival function, which is defined for a variable $Z = (Z_1, \ldots, Z_d)$ as

$$\overline{F}(x) = \mathbb{P}\{Z_1 \geq x_1, \ldots, Z_d \geq x_d\}. \tag{1.17}$$

1.2.3 The univariate case

Additional results can be obtained for the case $d = 1$. An immediate consequence of (1.16) is

$$1 - F(-x) = 2 F_0(x) - F(x), \qquad x \in \mathbb{R}, \tag{1.18}$$

which will be useful shortly.

The following representation can be obtained with an argument similar to Proposition 1.3. Note that $V = |Z|$ has distribution $2 f_0(\cdot)$ on $[0, \infty)$, irrespective of the modulation factor, and is of type (1.2). See Problem 1.2.

Proposition 1.6 *If Z_0 is a univariate variable having density f_0 symmetric about 0, $V = |Z_0|$ and G satisfies (1.4), then*

$$Z = S_V\, V, \qquad S_V = \begin{cases} +1 & \text{with probability } G(V), \\ -1 & \text{with probability } G(-V) \end{cases} \tag{1.19}$$

has density function (1.3).

We know that $\mathbb{E}\{Z^m\} = \mathbb{E}\{Z_0^m\} = \mathbb{E}\{V^m\}$ for $m = 0, 2, 4 \ldots$ The odd moments of Z can be expressed with the aid of (1.19) as

$$
\begin{aligned}
\mathbb{E}\{Z^m\} &= \mathbb{E}\{S_V\, V^m\} \\
&= \mathbb{E}_V\{\mathbb{E}\{S_V|V\}\, V^m\} \\
&= \mathbb{E}\{[G(V) - G(-V)]V^m\} \\
&= \mathbb{E}\{[2\, G(V) - 1]V^m\} \\
&= 2\, \mathbb{E}\{V^m\, G(V)\} - \mathbb{E}\{V^m\}, \qquad m = 1, 3, \ldots
\end{aligned} \tag{1.20}
$$

Consider now a fixed base density f_0 and a set of modulating functions G_k, all satisfying (1.4). What can be said about the resulting perturbed versions of f_0? This broad question can be expanded in many directions. An especially interesting one, tackled by the next proposition, is to find which conditions on the G_k ensure that there exists an ordering on the distribution functions

$$F_k(x) = \int_{-\infty}^{x} 2 f_0(u)\, G_k(u)\, du, \tag{1.21}$$

since this fact implies a similar ordering of moments and quantiles. If the variables X_1 and X_2 have distribution functions F_1 and F_2, respectively, recall that X_2 is said to be stochastically larger than X_1, written $X_2 \geq_{st} X_1$, if $\mathbb{P}\{X_2 > x\} \geq \mathbb{P}\{X_1 > x\}$ for all x, or equivalently $F_1(x) \geq F_2(x)$. In this case we shall also say that X_1 is stochastically smaller than X_2, written $X_1 \leq_{st} X_2$. An introductory account of stochastic ordering is provided by Whitt (2006).

Proposition 1.7 *Consider functions G_1 and G_2 on $\mathbb{R}$ which satisfy condition (1.4) and additionally $G_2(x) \geq G_1(x)$ for all $x > 0$. Then distribution functions (1.21) satisfy*

$$F_1(x) \geq F_2(x), \qquad x \in \mathbb{R}. \qquad (1.22)$$

If $G_1(x) > G_2(x)$ for all x in some interval, (1.22) holds strictly for some x.

Proof Consider first $s \leq 0$ and notice that $G_1(x) \geq G_2(x)$ for all $x < s$. This clearly implies $F_1(s) \geq F_2(s)$. If $s > 0$, the same conclusion holds using (1.18) with $x = -s$. QED

To illustrate, consider variables Z_0, Z and $|Z_0|$ whose respective densities are: (i) $f_0(x)$, (ii) $2 f_0(x) G(x)$ with G continuous and $\frac{1}{2} < G(x) < 1$ for $x > 0$, and (iii) $2 f_0(x) I_{[0,\infty)}(x)$. They can all be viewed as instances of (1.3), recalling that the first distribution is associated with $G(x) \equiv \frac{1}{2}$ and the third one with $G(x) = I_{[0,\infty)}(x)$, both fulfilling (1.4). From Proposition 1.7 it follows that

$$Z_0 \leq_{st} Z \leq_{st} |Z_0| \qquad (1.23)$$

and correspondingly, for any increasing function $t(\cdot)$, we can write

$$\mathbb{E}\{t(Z_0)\} < \mathbb{E}\{t(Z)\} < \mathbb{E}\{t(|Z_0|)\}, \qquad (1.24)$$

provided these expectations exist. Here strict inequalities hold because of analogous inequalities for the corresponding G functions, which implies strict inequality for some x in (1.22). A case of special interest is when $t(x) = x^{2k-1}$, for $k = 1, 2, \ldots$, leading to ordering of odd moments. Another implication of stochastic ordering is that p-level quantiles of the three distributions are ordered similarly to expectations in (1.24), for any $0 < p < 1$.

We often adopt the form of (1.2), with pertaining conditions, and it is convenient to formulate a version of Proposition 1.7 for this case.

Corollary 1.8 *Consider $G_1(x) = G_0\{w_1(x)\}$ and $G_2(x) = G_0\{w_2(x)\}$, where G_0, w_1 and w_2 satisfy (1.1) and additionally G_0 is monotonically increasing. If $w_2(x) \geq w_1(x)$ for all $x > 0$, then (1.22) holds. If $w_1(x) > w_2(x)$*

for all x in some interval of the positive half-line, (1.22) holds strictly for some x.

A further specialization occurs when $w_j(\cdot)$ represents an instance of the linear form $w(x) = \alpha x$, where α is an arbitrary constant, leading to the form (quite popular in this stream of literature)

$$f(x; \alpha) = 2\, f_0(x)\, G_0(\alpha x), \qquad x \in \mathbb{R}, \qquad (1.25)$$

where of course f_0 and G_0 are as in Proposition 1.1.

Corollary 1.9 *If f_0 and G_0 are as in Proposition 1.1 with $d = 1$, the set of densities (1.25) indexed by the real parameter α have distribution functions stochastically ordered with α.*

1.2.4 Bibliographic notes

A simplified version of Proposition 1.1 for the linear case of type $w(x) = \alpha x$ when $d = 1$ has been presented by Azzalini (1985); the rest of that paper focuses on the skew-normal distribution, which is the theme of the next two chapters. A follow-up paper (Azzalini, 1986) included, in the restricted setting indicated, stochastic representations analogous to those presented in § 1.2.2 and § 1.2.3, and a statement (his Proposition 1) equivalent to modulation invariance. Azzalini and Capitanio (1999, Section 7) introduced a substantially more general result, which will be examined later in this chapter.

The present version of Proposition 1.1 is as given by Azzalini and Capitanio (2003); the matching formulation (1.3)–(1.4) was developed independently by Wang et al. (2004), who showed the essential equivalence of the two constructions. Both papers included the corresponding general forms of stochastic representation and perturbation invariance. Wang et al. (2004) included also Proposition 1.5, up to an inessential modification. An intermediate formulation of similar type, where f_0 is a density of elliptical type, has been presented by Genton and Loperfido (2005).

The content of § 1.2.3 is largely based on § 3.1 of Azzalini and Regoli (2012a), with some exceptions: Proposition 1.6 and (1.20) have been given by Azzalini (1986), the latter up to a simple extension; inequalities similar to (1.24) have been obtained by Umbach (2006) for the case of an odd function $t(\cdot)$ such that $t(x) > 0$ for $x > 0$.

1.3 Some broader formulations

1.3.1 Other conditioning mechanisms

We want to examine more general constructions than that of Proposition 1.1, by relaxing the conditions involved. At first sight this programme seems pointless, recalling that, by Proposition 1.5, the set of distributions already encompassed is the widest possible. Such explorations make sense when we fix in advance some of the components; quite commonly, we want to pre-select the base density f_0. With these restrictions, the statement of Proposition 1.5 is affected.

As a first extension to the setting of Proposition 1.1, we replace the component $G_0\{w(x)\}$ by $G_0\{\alpha_0 + w(x)\}$, where α_0 is some fixed but arbitrary real number. This variant is especially natural if one thinks of the linear case $\alpha_0 + \alpha x$, which has been examined by various authors. With the same notation and type of argument adopted in the proof of Proposition 1.1, it follows that

$$f(x) = f_0(x) \frac{G_0\{\alpha_0 + w(x)\}}{\mathbb{P}\{T < \alpha_0 + w(Z_0)\}} \qquad (1.26)$$

is a density function on $\mathbb{R}^d$. We shall commonly refer to this distribution as an *extended* version of the similar one without α_0.

Such a simple modification of the formulation has an important impact on the whole construction, unless of course $\alpha_0 = 0$. One effect is that the denominator of (1.26) must be computed afresh for any choice of components. This computation is feasible in closed form only in favourable cases, while an appealing aspect of (1.2) is to have a fixed $\frac{1}{2}$ here.

In addition, the associated stochastic representation is affected. If we now set

$$S_{Z_0} = \begin{cases} +1 & \text{if } T < \alpha_0 + w(Z_0), \\ -1 & \text{otherwise,} \end{cases} \qquad (1.27)$$

then the distribution of $Z = (Z_0 | S_{Z_0} = 1)$ turns out to be (1.26), arguing as in Proposition 1.3. However, a representation similar to (1.10) does not hold because now $G(x) = G_0\{\alpha_0 + w(x)\}$ does not satisfy (1.4). In turn, this removes the modulation invariance property (1.12).

In spite of the above limitations, there are good reasons to explore this direction further. Although an explicit computation of the denominator in (1.26) cannot be worked out in general, still it can be pursued in a set of practically important cases. In addition, strong motivations arise from applications to consider this construction, and even more elaborate ones. In this section we only sketch a few general aspects, since a fuller treatment is feasible only in some specific cases, partly for the reasons explained; these developments will take place in later chapters.

It is convenient to reframe the probability context in a slightly different, but eventually equivalent, manner. Consider a $(d+m)$-dimensional variable (Z_0, Z_1) with joint density $f_*(x_0, x_1)$ such that Z_0 has marginal density f_0 on $\mathbb{R}^d$ and Z_1 has marginal density f_1 on $\mathbb{R}^m$. For a fixed Borel set $C \in \mathbb{R}^m$ having positive probability, consider the distribution of $(Z_0|Z_1 \in C)$, that is

$$f(x) = \frac{\int_C f_*(x, z)\, dz}{\int_C f_1(z)\, dz} = f_0(x)\, \frac{\mathbb{P}\{Z_1 \in C|Z_0 = x\}}{\mathbb{P}\{Z_1 \in C\}} \tag{1.28}$$

for $x \in \mathbb{R}^d$; from the first equality we see that $f(x)$ integrates to 1. In the special case when Z_0 and Z_1 are independent, the final fraction in (1.28) reduces to 1, and $f = f_0$.

The appeal of (1.28) comes from its meaningful interpretation from the viewpoint of applied work: $f(x)$ represents the joint distribution of a set of quantities of interest, Z_0, which are observed only for cases fulfilling a certain condition, that is $Z_1 \in C$, determined by another set of variables. As a simple illustration, think of Z_0 as the set of scores obtained by a student in certain university exams, and of Z_1 as the score(s) obtained by the same student in university admission test(s); we can observe Z_0 only for students whose Z_1 belongs to the admission set C. Situations of this type usually go under the heading 'selective sampling' or similar terms; it is then quite natural to denote (1.28) a *selection distribution*.

Expression (1.2) can be obtained as a special case of (1.28) when $m = 1$, $C = (-\infty, 0]$ and $Z_1 = T - w(Z_0)$, where T is a variable with distribution function G_0, independent of Z_0, and conditions (1.1) hold. Clearly (1.28) encompasses much more general situations, of which (1.26) is a subset. The next example is provided by two-sided constraints of the form $a < Z_1 < b$, again when $m = 1$. A much wider scenario is opened up by consideration of multiple constraints when $m > 1$.

Some general conclusions can be drawn about distributions of type (1.28). One of these is that, if Z_0 is transformed to $t(Z_0)$, the conditional distribution of $(t(Z_0)|Z_1 \in C)$ is still computed using (1.28), replacing the distribution of Z_0 with that of $t(Z_0)$. One implication is that, if f_0 belongs to a parametric family closed under a set of invertible transformations $t(\cdot)$, such as the set of affine transformations, then the same closure property holds for (1.28). See also Problem 1.8.

Because of its ample generality, it is difficult to develop more general conclusions for (1.28). As already indicated, in later chapters we shall examine important subcases, in particular those which allow a manageable computation of the two integrals involved, in connection with a symmetric density f_0, usually of elliptical type. The case of interest here is $m > 1$ since the case with $m = 1$ falls under the umbrella of the modulation invariance property.

Bibliographic notes

Emphasis has been placed on distributions of type (1.26), especially when $w(\cdot)$ is linear, by Barry Arnold and co-workers in a series of papers, many of which are summarized in Arnold and Beaver (2002); some will be described specifically later on. An initial formulation of (1.28) has been presented by Arellano-Valle *et al.* (2002), referring to the case where C is an orthant of $\mathbb{R}^m$, extended first by Arellano-Valle and del Pino (2004) and subsequently by Arellano-Valle and Genton (2005) and Arellano-Valle *et al.* (2006). The last paper shows how (1.28) formally encompasses a range of specific families of distributions examined in the literature. The focus on their development lies in situations where f_0 in (1.28) is a symmetric density; this case gives rise to what they denote *fundamental skew-symmetric* (FUSS) distributions. As already remarked, a unified theory does not appear to be feasible much beyond this point and specific, although very wide, subclasses must be examined. Some general results, however, have been provided by Arellano-Valle and Genton (2010a) with special emphasis on the distribution of quadratic forms when the parent population before selection has a normal or an elliptically contoured distribution.

1.3.2 Working with generalized symmetry

Proposition 1.10 *Denote by T a continuous real-valued random variable with distribution function G_0 symmetric about 0 and by Z_0 a d-dimensional variable with density function f_0, independent of T, such that the real-valued variable $W = w(Z_0)$ is symmetric about 0. Then*

$$f(x) = 2 f_0(x) G_0\{w(x)\}, \qquad x \in \mathbb{R}^d, \tag{1.29}$$

is a density function.

Proof See the final line of the 'instructive proof' of Proposition 1.1.   QED

Proposition 1.1 can be seen as a restricted version of this result, since conditions (1.1) are sufficient to ensure that $w(Z_0)$ has a symmetric distribution about 0. From an operational viewpoint the formulation in Proposition 1.1 is more convenient because checking conditions (1.1) is immediate, but does not embrace all possible settings falling within Proposition 1.10. Notice that Proposition 1.10 does not require that f_0 is symmetric about 0.

For a simple illustration, consider the density function on $\mathbb{R}^2$ obtained by modulating the bivariate normal with standardized marginals and correlations ρ, denoted $\varphi_B(x_1, x_2; \rho)$, as follows:

$$f(x) = 2\,\varphi_B(x_1, x_2; \rho)\,\Phi\{\alpha(x_1^2 - x_2^2)\}, \qquad x = (x_1, x_2) \in \mathbb{R}^2, \qquad (1.30)$$

where α is a real parameter and Φ is the standard normal distribution function. In this case the perturbation factor modifies the base density, preserving central symmetry. Figure 1.2 shows two instances of this density.

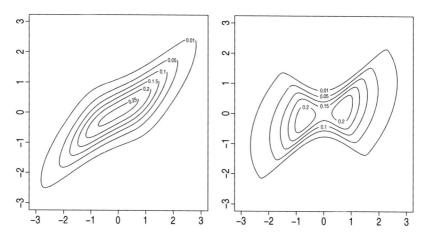

Figure 1.2 Density functions of type (1.30), displayed as contour level plots: in the left panel $\alpha = 1, \rho = 0.8$; in the right panel $\alpha = 3, \rho = 0.4$.

The fact that $f(x)$ integrates to 1 does not follow from Proposition 1.1 which requires an odd function $w(x)$, while $w(x) = \alpha(x_1^2 - x_2^2)$ is even; equivalently, $G(x) = \Phi\{\alpha(x_1^2 - x_2^2)\}$ does not satisfy (1.4). However, if $Z_0 = (Z_{01}, Z_{02})^\top \sim N_2(0, \Omega)$ where Ω is the 2×2 correlation matrix with off-diagonal entries ρ, it is true that $w(Z_0) = \alpha(Z_{01}^2 - Z_{02}^2)$ has a symmetric distribution about 0, and so Proposition 1.10 can be applied to conclude that (1.30) integrates to 1. In this respect, it would be irrelevant to replace Φ in (1.30) by some other symmetric distribution function G_0.

From the argument of the proof, it is immediate that a random variable with distribution (1.29) admits a representation of type (1.9). For the reasons already discussed in connection with (1.26), it is desirable that a representation similar to (1.10) also exists. The next result provides a set of sufficient conditions to this end.

Proposition 1.11 *Let T and Z_0 be as in Proposition 1.10, and suppose that there exists an invertible transformation $R(\cdot)$ such that, for all $x \in \mathbb{R}^d$,*

$$f_0(x) = f_0[R(x)], \quad |\det R'(x)| = 1, \quad w[R(x)] = -w(x), \qquad (1.31)$$

where $R'(x)$ denotes the Jacobian matrix of the partial derivatives, then

$$Z = \begin{cases} Z_0 & \text{if } T \le w(Z_0), \\ R^{-1}(Z_0) & \text{otherwise} \end{cases} \qquad (1.32)$$

has distribution (1.29).

Proof The density function of Z at x is

$$\begin{aligned} f(x) &= f_0(x)\, G_0\{w(x)\} + f_0(R(x))\, |\det R'(x)|\, [1 - G_0\{w(R(x))\}] \\ &= f_0(x)\, G_0\{w(x)\} + f_0(x)\, [1 - G_0\{-w(x)\}] \\ &= 2\, f_0(x)\, G_0\{w(x)\} \end{aligned}$$

using (1.31) and $G_0(-x) = 1 - G_0(x)$. QED

In this formulation the condition of (central) symmetry $f_0(x) = f_0(-x)$ has been replaced by the first requirement in (1.31), $f_0(x) = f_0[R(x)]$, which represents a form of *generalized symmetry*. Usual symmetry is recovered when $R(x) = -x$. The requirement of an odd function w is replaced here by the similarly generalized condition given by the last expression in (1.31).

For the corresponding extension of the modulation invariance property (1.12), consider a transformation from $\mathbb{R}^d$ to $\mathbb{R}^q$ which is even in the generalized sense adopted here, that is

$$t(x) = t(R^{-1}(x)), \qquad x \in \mathbb{R}^d.$$

It is immediate from representation (1.32) that (1.12) then holds.

For distribution (1.30), conditions (1.31) are fulfilled by the transformation

$$R(x) = R_0\, x, \qquad R_0 = \begin{pmatrix} 0 & 1 \\ 1 & 0 \end{pmatrix} = R_0^{-1},$$

which swaps the two coordinates, and $w(x) = \alpha(x_1^2 - x_2^2)$. Therefore, if $Z = (Z_1, Z_2)$ has density (1.30), perturbation invariance holds for any transformation $t(Z)$ such that $t((Z_1, Z_2)) = t((Z_2, Z_1))$. One implication is that

$Z^\top \Omega^{-1} Z \sim \chi_2^2$. Another consequence is that, since $t(x) = x_1 x_2 = x_2 x_1 = t(R_0 x)$, then $\mathbb{E}\{Z_1 Z_2\} = \rho$. Since central symmetry holds for $f(x)$, then $\mathbb{E}\{Z_1\} = \mathbb{E}\{Z_2\} = 0$ and so $\mathrm{cov}\{Z_1, Z_2\} = \rho$.

Using Proposition 1.10, one can construct distributions also with non-symmetric base density; see Problem 5.17 for an illustration.

Finally, note that the statement of Proposition 1.10 is still valid under somewhat weaker assumptions, as follows. We can relax the assumption about absolute continuity of all distributions involved, and allow G or the distribution of $w(Z_0)$ to be of discrete or mixed type, provided the condition $\mathbb{P}\{T - W(Z_0) \le 0\} = \frac{1}{2}$ in (1.29) still holds. A sufficient condition to meet this requirement is that at least one of T and $W(Z_0)$ is continuous.

Bibliographic notes

Proposition 1.10 has been presented by Azzalini and Capitanio (1999, Section 7). Although it was followed by a remark that the base density does not need to be symmetric, the ensuing development focused on elliptical distributions, and this route was followed in a number of subsequent papers, including extensions to the weaker condition of central symmetry; these have been quoted in earlier sections. The broader meaning of Proposition 1.10 has been reconsidered by Azzalini (2012), on which this section is based. Since exploration of this direction started only recently, no further discussion along this line will take place in the following chapters.

1.4 Complements

Complement 1.1 (Random number generation)   For sampling from distribution (1.2), both (1.9) and (1.10) provide a suitable technique for random number generation. However, in practice the first one is not convenient, since it involves rejection of half of the sampled Z_0's, on average.

To generate S_{Z_0}, both forms (1.11) are suitable. Which of the two variants is computationally more convenient depends on the specific instance under consideration. The second form involves computation of $G(x)$, which in practice is expressed as $G_0\{w(x)\}$. Since evaluation of $w(\cdot)$ is required in both cases, the comparison is then between computation of G_0 and generation of U versus generation of T. A general statement on which route is preferable is not possible, because the comparison depends on a number of factors, including the computing environment in use.

Further stochastic representations may exist for specific subclasses of (1.2), to be discussed in subsequent chapters. In these cases, they provide additional generation algorithms for random number generation.

Sampling from a distribution of type (1.26) is a somewhat different problem compared with (1.2), because only representation following (1.27) holds in general here. Its use implies rejection of a fraction of the sampled Z_0's, and the acceptance fraction can be as low as 0 if α_0 approaches $-\infty$. The more general set of distributions (1.28) can be handled in a similar manner: sample values (Z_0, Z_1) are drawn from f_*, and we accept only those Z_0's such that $Z_1 \in C$. For both situations, the problem of non-constant, and possibly very low, acceptance rate can be circumvented for specific subclasses of (1.26) or of (1.28) which allow additional stochastic representations that do not involve an acceptance–rejection technique; again, these will be discussed in subsequent chapters.

Complement 1.2 (A characterization)   The property of modulation invariance (1.12) leads to a number of corollaries for distributions of type (1.3) which share the same base density f_0; some of these corollaries appear in the next proposition. However, the interesting fact is not their isolated validity, but instead the fact that they are equivalent to each other and to representation (1.3), hence providing a characterization result.

More explicitly, if modulation invariance holds for all even $t(\cdot)$, this implies that the underlying distributions allow a representation of type (1.3) with common base f_0.

Proposition 1.12   *Consider variables $Z = (Z_1, ..., Z_d)^\top$ and $Y = (Y_1, ..., Y_d)^\top$ with distribution functions F and H, and density functions f and h, respectively; denote by $\overline{F}$ and $\overline{H}$ the survival functions of Z and Y, respectively, defined as in (1.17). The following conditions are then equivalent:*

   (a) densities $f(x)$ and $h(x)$ admit a representation of type (1.3) with the same symmetric base density $f_0(x)$;

   (b) $t(X) \stackrel{d}{=} t(Y)$, for any even q-dimensional function t on $\mathbb{R}^d$;

   (c) $\mathbb{P}\{Z \in A\} = \mathbb{P}\{Y \in A\}$, for any symmetric set $A \subset \mathbb{R}^d$;

   (d) $F(x) + \overline{F}(-x) = H(x) + \overline{H}(-x)$;

   (e) $f(x) + f(-x) = h(x) + h(-x)$ (a.e.).

Proof

(a)$\Rightarrow$(b)   This follows from the perturbation invariance property of Proposition 1.4.

(b)$\Rightarrow$(c)   Simply note that the indicator function of a symmetric set A is an even function.

(c)⇒(d) On setting

$$A_+ = \{s = (s_1, \dots, s_d) \in \mathbb{R}^d : s_j \le x_j, \forall j\},$$
$$A_- = \{s = (s_1, \dots, s_d) \in \mathbb{R}^d : -s_j \le x_j, \forall j\} = -A_+,$$
$$A_\cup = A_+ \cup A_- ,$$
$$A_\cap = A_+ \cap A_- ,$$

both $A_\cup$ and $A_\cap$ are symmetric sets. Hence we obtain:

$$
\begin{aligned}
F(x) + \overline{F}(-x) &= \mathbb{P}\{Z \in A_+\} + \mathbb{P}\{Z \in A_-\} \\
&= \mathbb{P}\{Z \in A_\cup\} + \mathbb{P}\{Z \in A_\cap\} \\
&= \mathbb{P}\{Y \in A_\cup\} + \mathbb{P}\{Y \in A_\cap\} \\
&= H(x) + \overline{H}(-x) .
\end{aligned}
$$

(d)⇒(e) Taking the dth mixed derivative of the final relationship in (d), relationship (e) follows.

(e)⇒(a) This follows from the representation given in Proposition 1.5.

QED

This proof is taken from Azzalini and Regoli (2012a). For the case $d = 1$, an essentially equivalent result has been given by Huang and Chen (2007, Theorem 1).

Complement 1.3 (On uniqueness of the mode) Another interesting theme concerns the range of possible shapes of the modulated density f, for a given base f_0. This is a very broad issue, only partly explored so far. A specific but important question is as follows: if f_0 has a unique mode, when does f also have a unique mode?

In the case $d = 1$, it is tempting to conjecture that a monotonic G preserves uniqueness of the mode of f_0, but this is dismissed by the example having $f_0 = \varphi$, N(0, 1) density and $G(x) = \Phi(x^3)$, where Φ is the N(0, 1) distribution function. Figure 1.3 illustrates graphically this case; the left panel displays G, the right panel shows f.

Sufficient conditions for uniqueness of the mode of f are given by the next statement, which we reproduce without proof from Azzalini and Regoli (2012a). Recall that *log-concavity* of a density means that the logarithm of the density is a concave function; in the univariate case, this property is equivalent to *strong unimodality* of the density (Dharmadhikari and Joag-dev, 1988, Theorem 1.10).

Proposition 1.13 *In case $d = 1$, if $G(x)$ in (1.3) is an increasing function*

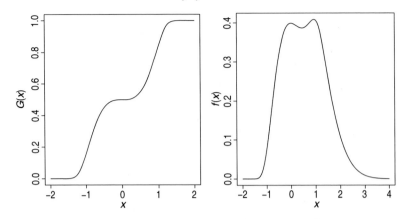

Figure 1.3 Example of a bimodal density produced with f_0 equal to the N(0, 1) density and $G(x) = \Phi(x^3)$; the left panel displays $G(x)$, the right panel the modulated density.

and $f_0(x)$ is unimodal at 0, then no negative mode exists. If we assume that f_0 and G have continuous derivatives everywhere on the support of f_0, $G(x)$ is concave for $x > 0$ and $f_0(x)$ is log-concave, where at least one of these properties holds in a strict sense, then there is a unique positive mode of $f(x)$. If $G(x)$ is decreasing, similar statements hold, with reversed sign of the mode; the uniqueness of the negative mode requires that $G(x)$ is convex for $x < 0$.

A popular situation where the conditions of this proposition are readily checked is (1.25) with linear w.

Corollary 1.14 *In case $d = 1$, if f_0 in Proposition 1.1 is log-concave and G'_0 is continuous everywhere and unimodal at 0, then density (1.25) is unimodal for all α, and the mode has the same sign as α.*

A related issue, which includes uniqueness of the mode as a byproduct, will be discussed in Chapter 6, for general d.

Complement 1.4 (Transformation of scale)   Jones (2013) has put forward an interesting proposal for the construction of flexible families of distributions which has a direct link with our main theme. We digress briefly in that direction for the aspects which illustrate this connection, without attempting a full summary of his formulation.

On the real line, consider a density f_0, symmetric about 0, having support

S_0. For a transformation t from the set S to $D \supseteq S_0$, it may happen that

$$f(x) = 2 f_0\{t(x)\}, \qquad x \in S \tag{1.33}$$

is a density function; in this case, the mechanism leading from f_0 to f is called *transformation of scale*, as opposed to the familiar transformation of variable. The next statement provides conditions to ensure that (1.33) is indeed a proper density.

Proposition 1.15 *Let $\bar{G} : D \to S$ denote a piecewise differentiable monotonically increasing function with inverse t, where $D \supseteq S_0 \ni 0$. If*

$$\bar{G}(z) - \bar{G}(-z) = z, \qquad \text{for all } z \in D \tag{1.34}$$

and f_0 is density symmetric about 0 with support S_0, then (1.33) is a density on S.

Proof Non-negativity of f follows from that of f_0, so we only need to prove that it integrates to 1. We consider the case where $\bar{G}$ is differentiable everywhere, with obvious extension to the case of piecewise differentiability. Making the substitution $z = t(x)$ and writing $G(z) = \bar{G}'(z)$, which is positive for all $z \in D$, write

$$\int_S 2\, f_0\{t(x)\}\, dx = 2 \int_D f_0(z)\, G(z)\, dz = 2 \int_{S_0} f_0(z)\, G(z)\, dz\,.$$

Since function G is positive and, on differentiating (1.34), fulfils conditions (1.4), then the above integral equals 1. QED

The argument of the proof shows that a variable X with distribution (1.33) can be obtained as $X = \bar{G}(Z)$, where Z has distribution of type (1.3) with $G = \bar{G}'$.

However, not all transformations of Z achieve the form (1.33), since $\bar{G}$ must satisfy (1.34). It can be shown that $t = \bar{G}^{-1}$ must be of the type $t(x) = x - s(x)$ where $s : \mathbb{R}^+ \to \mathbb{R}^+$ is an onto monotone decreasing function that is a self-inverse, i.e. $s^{-1}(x) = s(x)$. The proof of this fact is given by Jones (2013), together with various additional results. See also the related work of Jones (2012).

Complement 1.5 (Fechner-type distributions) A number of authors have considered asymmetric distributions on the real line obtained by applying different scale factors to the half-line $x > x_0$ and to the half-line $x < x_0$ of a density symmetric about x_0, which we can take equal to 0. This idea goes back to Fechner (1897, Chapter XIX) who applied it to the normal density, and it has re-emerged several times since then, in various forms of parameterization. See Mudholkar and Hutson (2000) for a variant form and an overview of others. Hansen (1994) employed the same device to build an asymmetric form of Student's distribution. A similar type of construction has been developed by Hinkley and Revankar (1977), by an independent argument, leading to a form of asymmetric Laplace distribution.

With similar logic, Arellano-Valle *et al.* (2005b) consider the class of densities

$$\frac{2}{a(\alpha) + b(\alpha)} \left[f_0\left(\frac{x}{a(\alpha)}\right) I_{[0,\infty)}(x) + f_0\left(\frac{x}{b(\alpha)}\right) I_{(-\infty,0)}(x) \right], \quad (1.35)$$

where f_0 is a density symmetric about 0 and α is a parameter which regulates asymmetry via the positive-valued functions $a(\cdot)$ and $b(\cdot)$. On setting $a(\alpha) = \alpha$ and $b(\alpha) = 1/\alpha$ where $\alpha > 0$, (1.35) leads to the class of Fernández and Steel (1998).

If X is a random variable with density (1.35), a stochastic representation is $X = W_\alpha |X_0|$ where X_0 has density $f_0(x)$ and W_α is an independent discrete variate such that

$$\mathbb{P}\{W_\alpha = a(\alpha)\} = \frac{a(\alpha)}{a(\alpha) + b(\alpha)}, \quad \mathbb{P}\{W_\alpha = -b(\alpha)\} = \frac{b(\alpha)}{a(\alpha) + b(\alpha)}.$$

Arellano-Valle *et al.* (2006) noted that this stochastic representation allows us to view (1.35) as an instance of the selection distributions (1.28). First note that $|X_0| \stackrel{d}{=} (X_0|X_0 > 0)$; hence set $X \stackrel{d}{=} (Z_0|Z_1 \in C)$ where $Z_0 = W_\alpha X_0$, $Z_1 = X_0$, $C = (0,\infty)$. Combining these settings, rewrite $X = W_\alpha |X_0|$ as $X = (W_\alpha X_0|X_0 > 0)$, which coincides with $X = (Z_0|Z_1 > 0)$.

Problems

1.1 Consider two independent real-valued continuous random variables, U and V, with common density f_0, symmetric about 0. Show that $Z_1 = \min\{U, V\}$ and $Z_2 = \max\{U, V\}$ have density of type (1.2) with base f_0.

1.2 Confirm that $V = |Z|$ introduced right before Proposition 1.6 has density $2 f_0(\cdot)$ on $[0,\infty)$ and find the expression of $G(x)$ to represent this distribution in the form (1.3).

1.3 Prove Proposition 1.6.

1.4 Assume that Z, conditionally on α, is a random variable with density function (1.25) and that α is a random variable with density symmetric about 0. Show that the unconditional density of Z is f_0. Extend this result to the general case (1.7) provided w is both an odd function of x for any fixed α and an odd function of α for any fixed x.

1.5 The product of two symmetric Beta densities rescaled to the interval $(-1, 1)$ takes the form

$$f_0(x, y) = \frac{(1 - x^2)^{a-1} \, (1 - y^2)^{b-1}}{4^{a+b-1} \, B(a, a) \, B(b, b)}, \qquad (x, y) \in (-1, 1)^2,$$

for some positive a and b. Define $f(x, y) = 2 \, f_0(x, y) \, L[w(x, y)]$, where $L(t) = (1 + \exp(-t))^{-1}$ is the standard logistic distribution function and

$$w(x, y) = \frac{\sin(p_1 x + p_2 y)}{1 + \cos(q_1 x + q_2 y)}.$$

Check that $f(x, y)$ is a properly normalized density on $(-1, 1)^2$. Choose constants $(a, b, p_1, p_2, q_1, q_2)$ as you like and plot the density using your favourite computing environment; repeat this step 11 more times.

1.6 For the variables in (1.23), show that $\mathrm{var}\{Z_0\} > \mathrm{var}\{Z\} > \mathrm{var}\{|Z_0|\}$, provided $\mathrm{var}\{Z_0\}$ exists.

1.7 Confirm that (1.26) is a density function.

1.8 Prove that, if a variable Z having selection distribution (1.28) with f_0 centrally symmetric is partitioned as $Z = (Z', Z'')$, then both the marginal distribution of Z' and that of Z' conditional on the value taken on by Z'' are still of the same type (Arellano-Valle and Genton, 2005).

1.9 Show that in (1.30) we can replace $w(x) = \alpha(x_1^2 - x_2^2)$ by

$$w(x) = \alpha_1(x_1 - x_2) + \cdots + \alpha_m(x_1^m - x_2^m)$$

for any natural number m and any choice of the coefficients $\alpha_1, \ldots, \alpha_m$, and still obtain a proper density function. Discuss the implication of selecting coefficients α_j where (i) only odd-order terms are non-zero, (ii) only even-order terms are non-zero (Azzalini, 2012).

1.10 If $\varphi_B(x_1, x_2; \rho)$ denotes the bivariate normal density with standardized marginals and correlation ρ, show that

$$2 \, \varphi_B(x_1, x_2; \rho) \, \Phi\{\alpha x_1 (x_2 - \rho x_1)\}, \quad 2 \, \varphi_B(x_1, x_2; \rho) \, \Phi\{\alpha x_2 (x_1 - \rho x_2)\},$$

for $(x_1, x_2) \in \mathbb{R}^2$, are density functions. Establish whether a representation of type (1.32) holds (Azzalini, 2012). *Note:* when $\rho = 0$, both forms reduce to a distribution examined by Arnold *et al.* (2002), which enjoys various interesting properties – its marginals are standardized normal densities and the conditional distribution of one component given the other is of skew-normal type, to be discussed in Chapter 2.

1.11 Consider the family of d-dimensional densities of type (1.2) where the base density is multivariate normal, $\varphi_d(x; \Sigma)$. Show that this family is closed under h-dimensional marginalization, for $1 \leq h < d$ (Lysenko *et al.*, 2009).

2

The skew-normal distribution: probability

Among the very many families of distributions which can be generated from (1.2), a natural direction to consider is some extension of the normal distribution, given its key role. This is the main target of the present chapter and the next one, which deal with the probability and the statistics side, respectively.

2.1 The basic formulation

2.1.1 Definition and first properties

If in (1.2) we select $f_0 = \varphi$ as the base density and $G_0 = \Phi$, the $N(0, 1)$ density function and distribution function, respectively, and $w(x) = \alpha x$, for some real value α, this produces the density function

$$\varphi(x; \alpha) = 2\,\varphi(x)\,\Phi(\alpha x) \qquad (-\infty < x < \infty), \tag{2.1}$$

whose graphical appearance is displayed in Figure 2.1 for a few choices of α. The integral function of $\varphi(x; \alpha)$ will be denoted $\Phi(x; \alpha)$.

For applied work we must introduce location and scale parameters. If Z is a continuous random variable with density function (2.1), then the variable

$$Y = \xi + \omega Z \qquad (\xi \in \mathbb{R}, \ \omega \in \mathbb{R}^+) \tag{2.2}$$

will be called a *skew-normal* (SN) variable with *location* parameter ξ, *scale* parameter ω, and *slant* parameter α. Its density function at $x \in \mathbb{R}$ is

$$\frac{2}{\omega}\,\varphi\!\left(\frac{x-\xi}{\omega}\right)\Phi\!\left(\alpha\,\frac{x-\xi}{\omega}\right) \equiv \frac{1}{\omega}\,\varphi\!\left(\frac{x-\xi}{\omega}; \alpha\right) \tag{2.3}$$

and we shall write

$$Y \sim \mathrm{SN}(\xi, \omega^2, \alpha),$$

where the square of ω is for analogy with the notation $N(\mu, \sigma^2)$. When

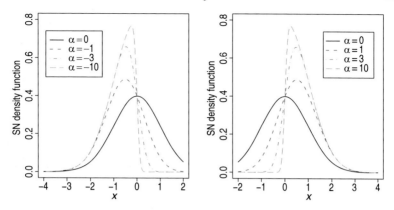

Figure 2.1 Skew-normal density functions when $\alpha = 0, -1, -3, -10$ in the left-hand panel, and $\alpha = 0, 1, 3, 10$ in the right-hand panel.

$\xi = 0, \omega = 1$, and we are back to density (2.1), we say that the distribution is 'normalized'. This is the case that will be considered most frequently in the present chapter.

There are various simple properties which follow immediately from the above definition and the general properties established in Section 1.2.

Proposition 2.1   *If Z denotes a random variable $SN(0, 1, \alpha)$, having density function $\varphi(x; \alpha)$, the following properties hold true:*

(a) $\varphi(x; 0) = \varphi(x)$ for all x;
(b) $\varphi(0; \alpha) = \varphi(0)$ for all α;
(c) $-Z \sim SN(0, 1, -\alpha)$, equivalently $\varphi(-x; \alpha) = \varphi(x; -\alpha)$ for all x;
(d) $\lim_{\alpha \to \infty} \varphi(x; \alpha) = 2 \varphi(x) I_{[0,\infty)}(x)$, for all $x \neq 0$;
(e) $Z^2 \sim \chi_1^2$, irrespective of α;
(f) if $Z' \sim SN(0, 1, \alpha')$ with $\alpha' < \alpha$, then $Z' <_{st} Z$.

Statement (e) follows from Proposition 1.4. The limit distribution (d) is called the χ_1 distribution and also the half-normal distribution. Stochastic ordering (f) is a special case of Corollary 1.9. In turn, (f) implies $\mathbb{E}\{Z'\} < \mathbb{E}\{Z\}$ and similar inequalities between quantiles of any level. The reflection property (c) is a special case of the more general fact stated in § 1.2.1.

The standard normal distribution is an element of the family of skew-normal densities, as indicated by property (a) above. For positive values of α we obtain a distribution skewed to the right, and for negative α a distribution skewed to the left. Another important connection with the normal family is the chi-square property (e). These facts, and additional ones to be presented later, support the adoption of the term *skew-normal* for this family.

2.1.2 Moment generating function and some implications

The following result on the normal distribution has been presented repeatedly in the literature, with or without proof; authors who have provided a proof include Ellison (1964) and Zacks (1981, pp. 53–54). Ellison's result is in fact more general; see Proposition B.1 on p. 233.

Lemma 2.2   *If $U \sim N(0, 1)$ then*

$$\mathbb{E}\{\Phi(h\,U + k)\} = \Phi\left(\frac{k}{\sqrt{1 + h^2}}\right), \qquad h, k \in \mathbb{R}. \tag{2.4}$$

From this result, the moment generating function of Y is readily obtained, that is

$$M(t) = \mathbb{E}\{\exp(\xi\,t + \omega\,Z\,t)\}$$
$$= 2\,\exp(\xi\,t + \tfrac{1}{2}\omega^2\,t^2) \int_{\mathbb{R}} \varphi(z - \omega t)\,\Phi(\alpha z)\,dz$$
$$= 2\,\exp(\xi\,t + \tfrac{1}{2}\omega^2\,t^2)\,\Phi(\delta\,\omega\,t) \tag{2.5}$$

where

$$\delta = \delta(\alpha) = \frac{\alpha}{\sqrt{1 + \alpha^2}}, \qquad \delta \in (-1, 1). \tag{2.6}$$

Multiplication of (2.5) by the moment generating function of the $N(\mu, \sigma^2)$ distribution, $\exp(\mu\,t + \sigma^2\,t^2/2)$, is still a function of type (2.5). After a simple reduction, we obtain the following statement.

Proposition 2.3   *If $Y_1 \sim SN(\xi, \omega^2, \alpha)$ and $Y_2 \sim N(\mu, \sigma^2)$ are independent random variables, then*

$$Y_1 + Y_2 \sim SN(\xi + \mu, \omega^2 + \sigma^2, \tilde{\alpha}), \qquad \tilde{\alpha} = \frac{\alpha}{\sqrt{1 + (1 + \alpha^2)\,\sigma^2/\omega^2}}. \tag{2.7}$$

In agreement with intuition, the slant parameter $\tilde{\alpha}$ of $Y_1 + Y_2$ is smaller in absolute value than the slant α of Y_1. Another point to notice is that

$$\lim_{\alpha \to \pm\infty} \tilde{\alpha} = \pm\frac{\omega}{\sigma}. \tag{2.8}$$

Consider now the case when Y_1 and Y_2 are both 'proper', that is with a non-null slant parameter. A natural question to ask is whether $Y_1 + Y_2$ is still of SN type. In other words, is the SN family closed under convolution? To proceed, we need the following preliminary result.

Lemma 2.4 *For any choice of the constants $a_1, b_1, a_2, b_2, c_0, c_1, c_2$ such that $b_1 \neq 0$ and $b_2 \neq 0$, there exist no constants a, b, d_0, d_1, d_2 such that*

$$\exp(c_0 + c_1 x + c_2 x^2)\,\Phi(a_1 + b_1 x)\,\Phi(a_2 + b_2\, x) = \exp(d_0 + d_1 x + d_2 x^2)\,\Phi(a + b\, x) \tag{2.9}$$

for all $x \in \mathbb{R}$.

Informal proof Denote by $h(x)$ the difference between the log-transformed left and right sides of (2.9), that is $h(x) = h_1(x) + h_2(x)$, where

$$h_1(x) = c_0 + c_1 x + c_2 x^2 - (d_0 + d_1 x + d_2 x^2),$$
$$h_2(x) = \log \Phi(a_1 + b_1 x) + \log \Phi(a_2 + b_2 x) - \log \Phi(a + bx),$$

such that $h(x) \equiv 0$ if (2.9) is true. Since h_1 is a polynomial and h_2 is a transcendental function, their sum is identically 0 only if both $h_1 \equiv 0$ and $h_2 \equiv 0$. If $h_2(x) = 0$ was true for all x, then after exponentiation this would imply that an equality of type (B.13) on p. 233 holds for all x, but this is ruled out by Proposition B.4. Hence $h_2 \not\equiv 0$ and so also $h \not\equiv 0$. QED

To address the above question of whether $Y_1 + Y_2$ is SN, notice first that a statement analogous to the above lemma holds removing a_1, a_2, a from (2.9). Under independence of the summands, the moment generating function of $Y_1 + Y_2$ has a form like the left-hand side of (2.9) with $c_2 > 0$, $b_1 b_2 \neq 0$. Since (2.9) cannot hold everywhere, this moment generating function does not have the form of the right-hand side of (2.9), that is of type (2.5). Hence we conclude that $Y_1 + Y_2$ is not of SN type.

An extension of Lemma 2.2 for SN variates can be obtained as follows. If $Z \sim SN(0, 1, \alpha)$ and $U \sim N(0, 1)$ are independent variables, then

$$\mathbb{E}\{\Phi(hZ + k)\} = \mathbb{E}\{\mathbb{P}\{U \leq h\,z + k | Z = z\}\}$$
$$= \mathbb{P}\{U - h\,Z \leq k\}$$

and, by applying Proposition 2.3 to the distribution of $U - hZ$, we arrive at the first statement below; the second one is obtained in a similar way.

Proposition 2.5 *If $Z \sim SN(0, 1, \alpha)$ and $U \sim N(0, 1)$, then*

$$\mathbb{E}\{\Phi(hZ + k)\} = \Phi\left(\frac{k}{\sqrt{1 + h^2}}; -\frac{h\,\alpha}{\sqrt{1 + h^2 + \alpha^2}}\right), \tag{2.10}$$

$$\mathbb{E}\{\Phi(hU + k; \alpha)\} = \Phi\left(\frac{k}{\sqrt{1 + h^2}}; \frac{\alpha}{\sqrt{1 + h^2(1 + \alpha^2)}}\right). \tag{2.11}$$

2.1.3 Stochastic representations

One of the more attractive features of the SN family is that it admits a variety of stochastic representations. These are useful for random number generation, and in some cases they provide a motivation for the adoption of the SN family as a stochastic model for observed data.

Conditioning and selective sampling

Recalling Proposition 1.3, a variable $Z \sim \mathrm{SN}(0, 1, \alpha)$ can be obtained by either of the representations

$$Z = \begin{cases} X_0 & \text{if } U < \alpha X_0, \\ -X_0 & \text{otherwise,} \end{cases} \qquad Z = (X_0 | U < \alpha X_0), \qquad (2.12)$$

where X_0 and U are independent $\mathrm{N}(0, 1)$ variables. For the purpose of pseudo-random number generation, the first variant is clearly more efficient, since it does not require any rejection, while the latter variant is more useful for theoretical considerations.

We can re-express this construction by introducing the bivariate normal variable (X_0, X_1) with standardized marginals where

$$X_1 = \frac{\alpha X_0 - U}{\sqrt{1 + \alpha^2}}$$

such that $\mathrm{cor}\{X_0, X_1\} = \delta(\alpha)$. Then representations (2.12) become

$$Z = \begin{cases} X_0 & \text{if } X_1 > 0, \\ -X_0 & \text{otherwise} \end{cases} \qquad Z = (X_0 | X_1 > 0). \qquad (2.13)$$

Although (2.13) is mathematically equivalent to the earlier construction based on (X_0, U), the second formulation has the advantage of an appealing interpretation from the point of view of stochastic modelling. In many practical cases, a variable X_0', say, is observed when another variable X_1', correlated with the first one, exceeds a certain threshold, leading to a situation of selective sampling. If this threshold corresponds to the mean value of X_1' and joint normality of (X_0', X_1') holds, we are effectively in case (2.13), up to an inessential change of location and scale between (X_0, X_1) and (X_0', X_1').

A natural remark is that in many cases the selection threshold is an arbitrary value, not the mean of X_1'. This more general case is connected to the variant form of the SN distribution to be discussed in Section 2.2.

Additive representation

Consider an arbitrary value $\delta \in (-1, 1)$ and use Proposition 2.3 in the limiting case (2.8) with $\omega = |\delta|$, $\sigma = \sqrt{1 - \delta^2}$ to obtain the next statement. If

U_0, U_1 are independent $N(0, 1)$ variates, then

$$Z = \sqrt{1 - \delta^2}\, U_0 + \delta\, |U_1| \sim SN(0, 1, \alpha) \qquad (2.14)$$

where

$$\alpha = \alpha(\delta) = \frac{\delta}{\sqrt{1 - \delta^2}}. \qquad (2.15)$$

Minima and maxima

Consider a bivariate normal random variable (X, Y) with standardized marginals and $\text{cor}\{X, Y\} = \rho$, and denote its density function by $\varphi_B(x, y; \rho)$. The distribution function of $Z_2 = \max\{X, Y\}$ is

$$H(t) = \mathbb{P}\{X \le t, Y \le t\}$$
$$= \int_{-\infty}^{t} \int_{-\infty}^{t} \varphi_B(x, y; \rho)\, dx\, dy$$
$$= \int_{-\infty}^{t} g(t, y)\, dy,$$

where $g(t, y) = \int_{-\infty}^{t} \varphi_B(x, y; \rho)\, dx$. The density function is

$$H'(t) = g(t, t) + \int_{-\infty}^{t} g_t'(t, y)\, dy$$
$$= 2\, g(t, t)$$
$$= 2 \int_{-\infty}^{t} \varphi(t)\, \frac{1}{\sqrt{1 - \rho^2}}\, \varphi\left(\frac{x - \rho t}{\sqrt{1 - \rho^2}} \right) dx$$
$$= 2\, \varphi(t) \int_{-\infty}^{\alpha t} \varphi(u)\, du,$$

where g_t' denotes the first partial derivative of g, that is $g_t'(t, y) = \varphi_B(t, y; \rho)$, and

$$\alpha = \sqrt{\frac{1 - \rho}{1 + \rho}}.$$

A similar computation holds for $Z_1 = \min\{X, Y\}$. We summarize the above discussion by writing

$$Z_1 \sim SN(0, 1, -\alpha), \qquad Z_2 \sim SN(0, 1, \alpha). \qquad (2.16)$$

2.1.4 Moments and other characteristic values

To compute the moments of $Y \sim \mathrm{SN}(\xi, \omega^2, \alpha)$, one route is via the moment generating function (2.5) or, equivalently but somewhat more conveniently, via the cumulant generating function

$$K(t) = \log M(t) = \xi t + \tfrac{1}{2}\omega^2 t^2 + \zeta_0(\delta \omega t) \qquad (2.17)$$

where

$$\zeta_0(x) = \log \{2 \, \Phi(x)\}. \qquad (2.18)$$

We shall also make use of the derivatives

$$\zeta_r(x) = \frac{\mathrm{d}^r}{\mathrm{d}x^r} \zeta_0(x) \qquad (r = 1, 2, \ldots) \qquad (2.19)$$

whose expressions, for the lower orders, are

$$
\begin{aligned}
\zeta_1(x) &= \varphi(x)/\Phi(x), \\
\zeta_2(x) &= -\zeta_1(x)\{x + \zeta_1(x)\} \\
&= -\zeta_1(x)^2 - x\,\zeta_1(x), \\
\zeta_3(x) &= -\zeta_2(x)\{x + \zeta_1(x)\} - \zeta_1(x)\{1 + \zeta_2(x)\} \\
&= 2\zeta_1(x)^3 + 3x\zeta_1(x)^2 + x^2\zeta_1(x) - \zeta_1(x), \\
\zeta_4(x) &= -\zeta_3(x)\{x + 2\zeta_1(x)\} - 2\zeta_2(x)\{1 + \zeta_2(x)\} \\
&= -6\,\zeta_1(x)^4 - 12\,x\,\zeta_1(x)^3 - 7\,x^2\,\zeta_1(x)^2 + 4\,\zeta_1(x)^2 \\
&\quad -x^3\,\zeta_1(x) + 3\,x\,\zeta_1(x),
\end{aligned}
\qquad (2.20)
$$

where $\zeta_1(x)$ coincides with the inverse Mills ratio evaluated at $-x$. All $\zeta_r(x)$ for $r > 1$ can be written as functions of $\zeta_1(x)$ and powers of x. For later use, notice that

$$\zeta_1(x) > 0, \qquad x + \zeta_1(x) > 0, \qquad \zeta_2(x) < 0, \qquad (2.21)$$

where the second inequality follows from a well-known property of Mills ratio; see (B.3) on p. 232.

Using (2.20), derivatives of $K(t)$ up to fourth order are immediate, leading to

$$\mathbb{E}\{Y\} = \xi + \omega\mu_z, \qquad (2.22)$$

$$\mathrm{var}\{Y\} = (\omega\,\sigma_z)^2, \qquad (2.23)$$

$$\mathbb{E}\{(Y - \mathbb{E}\{Y\})^3\} = \tfrac{1}{2}(4 - \pi)\,(\omega\mu_z)^3, \qquad (2.24)$$

$$\mathbb{E}\{(Y - \mathbb{E}\{Y\})^4\} = 2\,(\pi - 3)\,(\omega\mu_z)^4, \qquad (2.25)$$

where

$$\mu_z = \mathbb{E}\{Z\} = b\,\delta, \qquad \sigma_z^2 = \mathrm{var}\{Z\} = 1 - \mu_z^2 = 1 - b^2\delta^2 \qquad (2.26)$$

and

$$b = \zeta_1(0) = \sqrt{2/\pi}. \tag{2.27}$$

Standardization of the third and fourth cumulant produces the commonly used measures of skewness and kurtosis, that is

$$\gamma_1\{Y\} = \gamma_1\{Z\} = \frac{4 - \pi}{2} \frac{\mu_z^3}{\sigma_z^3}, \tag{2.28}$$

$$\gamma_2\{Y\} = \gamma_2\{Z\} = 2(\pi - 3) \frac{\mu_z^4}{\sigma_z^4}, \tag{2.29}$$

respectively. From the pattern of derivatives of $K(t)$ of order greater than two,

$$K^{(r)}(t) = (\delta\omega)^r \zeta_r(\omega\delta t), \qquad r > 2,$$

it is visible that the rth-order cumulant of Y is proportional to $(\delta\omega)^r$. Unfortunately, explicit computation of term $\zeta_r(0)$ does not seem feasible.

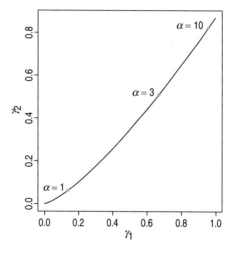

Figure 2.2 SN distribution: locus of γ_1 and γ_2 as α ranges from 0 to ∞, with labels corresponding to a few choices of α. When α takes on negative values, the curve is mirrored on the opposite side of the vertical axis.

Since γ_1 and γ_2 are often employed as measures of skewness and excess kurtosis, respectively, their behaviour and numerical range are of interest. Figure 2.2 shows graphically how γ_1 and γ_2 relate to each other and to α. From the above expressions, it is seen that they depend on the parameters only via μ_z/σ_z, which in turn increases monotonically with α up to

$b/\sqrt{1-b^2}$. Hence the ranges of γ_1 and γ_2 are

$$(-\gamma_1^{\max}, \gamma_1^{\max}), \qquad [0, \gamma_2^{\max}), \qquad (2.30)$$

respectively, where

$$\gamma_1^{\max} = \frac{\sqrt{2}(4-\pi)}{(\pi-2)^{3/2}} \approx 0.9953, \qquad \gamma_2^{\max} = \frac{8(\pi-3)}{(\pi-2)^2} \approx 0.8692. \qquad (2.31)$$

These ranges are not very wide, showing that the SN family does not provide an adequate stochastic model for cases with high skewness or kurtosis. Furthermore, one cannot choose γ_1 independently from γ_2, since they are both regulated by α.

A simple route to compute the nth moment of $Z \sim SN(0, 1, \alpha)$ is as follows; our real interest is to compute the odd moments. For any positive h, define

$$K_n(h) = \int_{-\infty}^{\infty} x^n \exp(-\tfrac{1}{2}h\,x^2)\,\Phi(x)\,dx$$

such that

$$K_0(h) = \left(\frac{\pi}{2h}\right)^{1/2}, \qquad K_1(h) = \frac{1}{h\sqrt{1+h}}$$

from the normalization factor of $\varphi(x;\alpha)$ and the expression of $\mathbb{E}\{Z\}$. Moreover, integration by parts lends the recurrence relationship

$$K_n(h) = -h^{-1} \int_{-\infty}^{\infty} (-h\,x)\,\exp(-\tfrac{1}{2}h\,x^2)\,x^{n-1}\,\Phi(x)\,dx$$

$$= h^{-1} \int_{-\infty}^{\infty} \exp(-\tfrac{1}{2}h\,x^2)\big((n-1)x^{n-2}\,\Phi(x) + x^{n-1}\,\varphi(x)\big)\,dx$$

$$= \frac{n-1}{h}K_{n-2}(h) + \frac{v_{n-1}}{h\,(1+h)^{n/2}}, \qquad n = 2, 3, \ldots$$

where v_k is the kth moment of the $N(0, 1)$ distribution, that is,

$$v_k = \begin{cases} 0 & \text{if } k = 1, 3, 5, \ldots \\ (k-1)!! & \text{if } k = 2, 4, 6, \ldots \end{cases}$$

from (B.9) on p. 233. It now follows that, if $\alpha \neq 0$,

$$\mathbb{E}\{Z^n\} = \int_{-\infty}^{\infty} x^n\,2\,\varphi(x)\,\Phi(\alpha x)\,dx$$

$$= \frac{2}{\alpha} \int_{-\operatorname{sgn}(\alpha)\infty}^{\operatorname{sgn}(\alpha)\infty} (\alpha^{-1}\,t)^n \varphi(\alpha^{-1}\,t)\,\Phi(t)\,dt$$

$$= \sqrt{\frac{2}{\pi}}\,\frac{\operatorname{sgn}(\alpha)}{\alpha^{n+1}}\,K_n(\alpha^{-2}), \qquad n = 0, 1, 2, \ldots \qquad (2.32)$$

Similarly to the mean value, the median and other quantiles are also increasing functions of α, thanks to Proposition 2.1(f). As for the mode, we first need to establish the following fact.

Proposition 2.6 *The distribution* $\mathrm{SN}(\xi, \omega^2, \alpha)$ *is log-concave, that is, the logarithm of its density is a concave function.*

Proof It suffices to prove the statement for the case $\mathrm{SN}(0, 1, \alpha)$, since the property is not altered by a change of location and scale. Taking into account (2.20) and the second inequality of (2.21), we get

$$\frac{d^2}{dx^2} \log \varphi(x; \alpha) = -1 - \zeta_1(\alpha x) \alpha^2 \{\alpha x + \zeta_1(\alpha x)\} < 0. \qquad \text{QED}$$

Hence the mode is unique. In fact, the more stringent result of strong unimodality holds since, in the univariate case, this property coincides with log-concavity of a distribution. Denote by $m_0(\alpha)$ the mode of $\mathrm{SN}(0, 1, \alpha)$; in the general case, the mode is $\xi + \omega \, m_0(\alpha)$. A somewhat peculiar feature is that $m_0(\alpha)$ has non-monotonic behaviour, since $m_0(0) = m_0(\infty) = 0$. For general α, no explicit expression of $m_0(\alpha)$ is available, and it must be evaluated by numerical maximization. A simple but practically quite accurate approximation is

$$m_0(\alpha) \approx \mu_z - \frac{\gamma_1 \, \sigma_z}{2} - \frac{\mathrm{sgn}(\alpha)}{2} \exp\left(-\frac{2\pi}{|\alpha|}\right), \qquad (2.33)$$

where the first two terms are obtained by the widely applicable approximation given, for instance, by Cramér (1946, p. 184) and the last term is a refinement applied to this specific distribution, obtained by numerical interpolation of the exact values.

Figure 2.3 displays the behaviour of mean, median and mode as functions of α (left panel) and δ (right panel). Only positive values of the parameters have been considered, because of Proposition 2.1(c). The maximal value of the mode occurs at $\alpha \approx 1.548$, $\delta \approx 0.8399$, where its value is about 0.5427. Use of (2.33) would reproduce quite closely the exact mode plotted in Figure 2.3.

2.1.5 Distribution function and tail behaviour

Consider the distribution function $\Phi(x; \alpha)$ of the $\mathrm{SN}(0, 1, \alpha)$ density. A straightforward computation gives

$$\Phi(x; \alpha) = 2 \int_{-\infty}^{x} \int_{-\infty}^{\alpha t} \varphi(t) \, \varphi(u) \, du \, dt$$

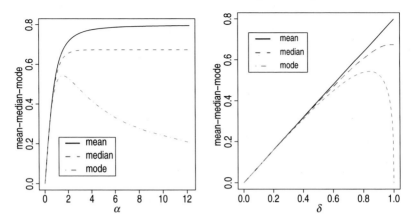

Figure 2.3 Mean, median and mode for SN distribution as a function of α (left panel) and δ (right panel) when $\alpha > 0$.

$$= 2 \int_{-\infty}^{x} \int_{-\infty}^{0} \varphi(t)\, \varphi\left(\frac{w + \delta t}{\sqrt{1 - \delta^2}}\right) \frac{1}{\sqrt{1 - \delta^2}}\, dw\, dt \qquad (2.34)$$

$$= 2\Phi_B(x, 0; -\delta), \qquad (2.35)$$

where $\Phi_B(x, y; \rho)$ is the standard bivariate normal distribution function.

It is convenient to express $\Phi(x; \alpha)$ in an alternative way based on the function

$$T(h, a) = \frac{1}{2\pi} \int_{0}^{a} \frac{\exp\{-\frac{1}{2}h^2(1 + x^2)\}}{1 + x^2}\, dx, \qquad h, a \in \mathbb{R}, \qquad (2.36)$$

studied by Owen (1956) in connection with the bivariate normal integral. Using the relationship between the bivariate normal integral and $T(h, a)$, and its properties recalled in Appendix B, we can rewrite (2.35) as

$$\Phi(x; \alpha) = \Phi(x) - 2\,T(x, \alpha). \qquad (2.37)$$

Computation of $\Phi(x; \alpha)$ becomes therefore quite manageable, since there exist efficient numerical methods for evaluating $T(h, a)$.

Proposition 2.7 *The following properties of $\Phi(x; \alpha)$ hold:*

(a) $\Phi(-x; \alpha) = 1 - \Phi(x; -\alpha),$

(b) $\Phi(x; \alpha) = 2\,\Phi(x)\,\Phi(\alpha x) + \begin{cases} 1 - \Phi(\alpha x; 1/\alpha) & \text{if } \alpha < 0, \\ 0 & \text{if } \alpha = 0, \\ -\Phi(\alpha x; 1/\alpha) & \text{if } \alpha > 0, \end{cases}$

(c) $\Phi(x; 1) = \{\Phi(x)\}^2,$

(d) $\Phi(0; \alpha) = \dfrac{1}{2} - \dfrac{\arctan \alpha}{\pi} = \dfrac{\arccos \delta(\alpha)}{\pi},$

(e) $\sup_x |\Phi(x; \alpha) - \Phi(x)| = \dfrac{\arctan |\alpha|}{\pi}.$

Property (a) follows from Proposition 2.1(c). Property (b) follows from the integration

$$\int \varphi(x)\,\Phi(\alpha x)\,dx = \Phi(x)\,\Phi(\alpha x) - \alpha \int \Phi(x)\,\varphi(\alpha x)\,dx$$

by parts. Setting $\alpha = 1$ in (b), one obtains (c). Expression (d) can be obtained by setting $x = 0$ in (2.35) and making use for the expression for the quadrant probability of a standardized bivariate normal variable; see (B.17) on p. 234. Finally, (e) follows from

$$\sup_x |\Phi(x; \alpha) - \Phi(x)| = 2\,\sup_x |T(x, \alpha)| = 2\,|T(0, \alpha)| = \dfrac{|\arctan \alpha|}{\pi}$$

taking into account that $|T(h, a)|$ is a decreasing function of h^2, as is clear from (2.36), except for the special case $a = 0$ where $T(h, 0) \equiv 0$.

The right and left tail probabilities of a skew-normal distribution have different rates of decay to zero. Property (a) of Proposition 2.7 allows us to examine the problem assuming $x > 0$. We then consider $1 - \Phi(x; \alpha)$ for the two cases $\alpha > 0$ and $\alpha < 0$ separately, and $x > 0$. If $\alpha = 0$, recall the classical result (B.3) for the normal distribution.

Proposition 2.8

$$\lim_{x \to +\infty} \dfrac{1 - \Phi(x; \alpha)}{2\,x^{-1}\,\varphi(x)} = 1 \quad \text{if } \alpha > 0, \qquad \lim_{x \to +\infty} \dfrac{1 - \Phi(x; \alpha)}{q(x, \alpha)} = 1 \quad \text{if } \alpha < 0$$

where

$$q(x, \alpha) = \sqrt{\dfrac{2}{\pi}}\,\dfrac{\varphi(x\,\sqrt{1 + \alpha^2})}{|\alpha|\,(1 + \alpha^2)\,x^2}. \tag{2.38}$$

For a sketch of the proof and more details, see Complement 2.4.

2.2 Extended skew-normal distribution

2.2.1 Introduction and basic properties

Lemma 2.2 prompts the introduction of an extension of the SN family of distributions, since

$$\dfrac{1}{\Phi(\alpha_0/\sqrt{1 + \alpha^2})} \int_{-\infty}^{\infty} \varphi(x)\,\Phi(\alpha_0 + \alpha x)\,dx = 1$$

for any choice of α_0, α. It is equivalent to adopt a simple modification of the parameters, and to consider the density function

$$\varphi(x; \alpha, \tau) = \varphi(x) \frac{\Phi(\tau \sqrt{1 + \alpha^2} + \alpha x)}{\Phi(\tau)}, \qquad x \in \mathbb{R}, \qquad (2.39)$$

where $(\alpha, \tau) \in \mathbb{R} \times \mathbb{R}$.

Since (2.39) reduces to (2.1) when $\tau = 0$, this explains the addition of the term 'extended' for this distribution, and more generally for any variable of type $Y = \xi + \omega Z$, if Z has density function of type (2.39). We shall write $Y \sim \mathrm{SN}(\xi, \omega^2, \alpha, \tau)$, where the presence of the component τ indicates that we are referring to an 'extended SN' distribution, briefly ESN. Notice that the value of τ becomes irrelevant when $\alpha = 0$. It is immediate that

$$-Y \sim \mathrm{SN}(-\xi, \omega^2, -\alpha, \tau).$$

Figure 2.4 displays the shape of the density (2.39) for α equal to 3 (in the left panel) or 10 (right panel) and a few choices of τ. It is visible that the effect of the new parameter τ is not independent of α. For the smaller choice of α, the effect of varying τ from the baseline value $\tau = 0$ is much the same as could be achieved by retaining $\tau = 0$ and selecting a suitable value of α. For $\alpha = 10$, the variation of τ modifies the density function in a more elaborate way.

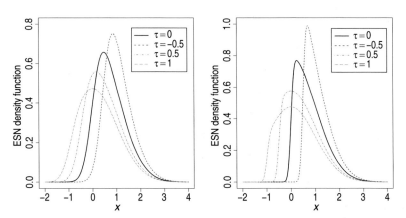

Figure 2.4 Extended skew-normal density functions when $\alpha = 3$ and $\tau = 0, -0.5, 0.5, 1$ in the left-hand panel and $\alpha = 10$ with the same values of τ in the right-hand panel.

Computation of the moment generating function of $Y = \xi + \omega Z$ where $Z \sim \mathrm{SN}(0, 1, \alpha, \tau)$ is very much the same as the SN case. Making use of

Lemma 2.2 again, one arrives at

$$
\begin{aligned}
M(t) &= \mathbb{E}\{\exp(\xi\, t + \omega\, Z\, t)\} \\
&= \frac{1}{\Phi(\tau)}\, \exp(\xi\, t + \tfrac{1}{2}\omega^2\, t^2) \int_{\mathbb{R}} \varphi(z - \omega t)\, \Phi\left(\tau\sqrt{1+\alpha^2} + \alpha z\right) dz \\
&= \exp\left(\xi\, t + \tfrac{1}{2}\omega^2\, t^2\right) \frac{\Phi(\tau + \delta\,\omega\, t)}{\Phi(\tau)},
\end{aligned}
\tag{2.40}
$$

where δ is again given by (2.6).

The similarity of the ESN and the SN moment generating functions implies that many other facts proceed in parallel for the two families. An example is the next statement, which matches closely Proposition 2.3.

Proposition 2.9 *If $Y \sim \mathrm{SN}(\xi, \omega^2, \alpha, \tau)$ and $U \sim \mathrm{N}(\mu, \sigma^2)$ are independent random variables, then $Y + U \sim \mathrm{SN}(\xi + \mu, \omega^2 + \sigma^2, \tilde{\alpha}, \tau)$, where $\tilde{\alpha}$ is given by (2.7).*

There is, however, one important aspect in which the SN and ESN families differ. Since the ESN density is not of the form (1.2), properties linked to Proposition 1.4 do not hold. A noteworthy consequence is the lack of a chi-square property similar to Proposition 2.1(e).

2.2.2 Stochastic representations

Selective sampling

Similarly to (2.13), denote by (X_0, X_1) a bivariate normal variable with standardized components and correlation δ, but with more general threshold of the type $X_1 + \tau > 0$ for an arbitrary value τ. A direct computation then shows

$$
Z = (X_0 | X_1 + \tau > 0) \sim \mathrm{SN}(0, 1, \alpha(\delta), \tau),
\tag{2.41}
$$

where $\alpha(\delta)$ is given by (2.15). In fact, if $\varphi_B(x, y; \delta)$ denotes the density of (X_0, X_1), then the density function of Z is

$$
\begin{aligned}
f_Z(x) &= \frac{1}{1 - \Phi(-\tau)} \int_{-\tau}^{\infty} \varphi_B(x, y; \delta)\, dy \\
&= \frac{1}{\Phi(\tau)} \int_{-\tau}^{\infty} \varphi(x)\, \varphi\left(\frac{y - \delta x}{\sqrt{1 - \delta^2}}\right) \frac{1}{\sqrt{1 - \delta^2}}\, dy \\
&= \varphi(x)\, \Phi(\tau)^{-1} \Phi(\tau\sqrt{1+\alpha^2} + \alpha x),
\end{aligned}
\tag{2.42}
$$

which confirms the above statement.

Representation (2.41) provides a probabilistic interpretation of the fact noted shortly after (2.39) that τ becomes irrelevant when $\alpha = 0$. In this case $\delta(\alpha) = 0$, so that X_0 and X_1 are independent, and the conditioning in (2.41) becomes void in probability, for any choice of τ.

Additive representation

Representation (2.14) can be extended to

$$Z = \sqrt{1 - \delta^2}\, U_0 + \delta\, U_{1,-\tau} \ \sim\ \mathrm{SN}(0, 1, \alpha(\delta), \tau), \qquad (2.43)$$

where $\delta \in (-1, 1)$ and $U_{1,-\tau}$ is a variable with distribution $N(0, 1)$ truncated below $-\tau$. To see this fact, start with two independent $N(0, 1)$ variables, U_0 and U_1, say, and define

$$X_0 = \sqrt{1 - \delta^2}\, U_0 + \delta\, U_1, \quad X_1 = U_1$$

so that (X_0, X_1) has the same distribution required for (2.41) and the condition $X_1 + \tau > 0$ establishes the same event, leading to a left-truncated normal variable of the same type as $U_{1,-\tau}$. Hence, the random variables Z (2.41) and (2.43) have the same distribution.

The argument in the above paragraph highlights the close mathematical connection between the representations (2.41) and (2.43) and their SN counterparts (2.13) and (2.14), respectively. The third representation, namely (2.16), is not known to have a counterpart for the ESN case.

For generation of pseudo-random numbers there is some difference between the SN and ESN cases. While for the SN case (2.12) and (2.14) provide a computationally efficient mechanism, the ESN case is slightly less favourable. The representation via conditioning (2.41) suffers from the problem that the rejection rate of sampled X_0 values depends on τ, and this can be very small if τ is a large negative value. To avoid the problem of high rejection rate of the sampled values, it is preferable to resort to (2.43), sampling $U_{0,-\tau}$ from the truncated normal variable. This can be accomplished by sampling from a variable uniformly distributed in $(\Phi(-\tau), 1)$, followed by the transformation $\Phi^{-1}(\cdot)$.

2.2.3 Cumulants and other properties

From (2.40), the cumulant generating function is given by

$$K(t) = \log M(t) = \xi t + \tfrac{1}{2}\omega^2 t^2 + \zeta_0(\tau + \delta\omega t) - \zeta_0(\tau)$$

whose derivatives are

$$K'(t) = \xi + \omega^2 t + \zeta_1(\tau + \delta\omega t)\,\delta\omega,$$
$$K''(t) = \omega^2 + \zeta_2(\tau + \delta\omega t)\,\delta\,\omega,$$
$$K^{(r)}(t) = \zeta_r(\tau + \delta\omega t)\,\delta^r\,\omega^r, \qquad \text{for } r > 2,$$

taking into account (2.19). From here one obtains that

$$\mathbb{E}\{Y\} = \xi + \zeta_1(\tau)\omega\delta, \tag{2.44}$$

$$\text{var}\{Y\} = \omega^2\{1 + \zeta_2(\tau)\,\delta^2\}, \tag{2.45}$$

$$\gamma_1\{Y\} = \frac{\zeta_3(\tau)\,\delta^3}{(1 + \zeta_2(\tau)\,\delta^2)^{3/2}} = \gamma_1\{Z\}, \tag{2.46}$$

$$\gamma_2\{Y\} = \frac{\zeta_4(\tau)\,\delta^4}{(1 + \zeta_2(\tau)\,\delta^2)^2} = \gamma_2\{Z\}. \tag{2.47}$$

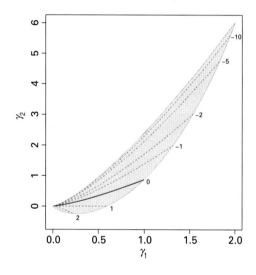

Figure 2.5 ESN distribution: locus of γ_1 and γ_2 as α ranges from 0 to ∞; the dashed lines represent the loci corresponding to some choices of τ, the solid line represents the locus corresponding to $\tau = 0$.

Figure 2.5 shows graphically the range of (γ_1, γ_2) with the ESN distribution with positive α; if α is negative, the plot is mirrored on the opposite side of the vertical axis. The dashed lines represent the loci corresponding to some choices of τ, indicated next to each line, as α varies in $(0, \infty)$; the line with $\tau = 0$ corresponds to Figure 2.2. The range of (γ_1, γ_2) is wider than in the SN case, and a small portion refers to negative γ_2 values, but on the whole the range is still limited. Qualitatively, this outcome was to be expected considering Figure 2.4, which does not display a major variation from Figure 2.1, plus the consideration that the tails are regulated by a mechanism very similar to the SN case.

Similarly to the SN case, the ESN distribution function can be expressed via the bivariate normal integral, because of (2.41). Specifically, using (2.42), the integral of (2.39) over $(-\infty, x)$ can be written as

$$\Phi(x; \alpha, \tau) = \frac{1}{\Phi(\tau)} \int_{-\infty}^{x} \int_{-\infty}^{\tau} \varphi(t) \, \varphi\left(\frac{u + \delta t}{\sqrt{1 - \delta^2}}\right) \frac{1}{\sqrt{1 - \delta^2}} \, du \, dt$$

$$= \frac{\Phi_B(x, \tau; -\delta)}{\Phi(\tau)}, \qquad (2.48)$$

where $\Phi_B(x, y; \rho)$ denotes the standard bivariate normal integral. From (B.21) in Appendix B, an alternative expression is

$$\Phi(x; \alpha, \tau) = \Phi(x) - \frac{1}{\Phi(\tau)} \Big[T\Big(x, \alpha + x^{-1}\tau \sqrt{1 + \alpha^2}\Big) - T(x, x^{-1}\tau)$$

$$+ T\Big(\tau, \alpha + \tau^{-1}x \sqrt{1 + \alpha^2}\Big) - T(\tau, \tau^{-1}x)\Big] \qquad (2.49)$$

which reduces to (2.37) when $\tau = 0$, taking into account (B.19).

Proposition 2.10   *If $W \sim SN(\xi, \omega^2, \alpha, \tau)$ and, conditionally on $W = w$, $Y \sim N(w, \sigma^2)$, then*

$$(W|Y = y) \sim SN(\xi_c, \omega_c^2, \alpha_c, \tau_c),$$

$$\xi_c = \frac{\sigma^{-2} y + \omega^{-2} \xi}{\sigma^{-2} + \omega^{-2}}, \quad \omega_c^2 = \frac{1}{\sigma^{-2} + \omega^{-2}}, \quad \alpha_c = \frac{\alpha}{\sqrt{1 + \omega^2/\sigma^2}},$$

$$\tau_c = \tau \sqrt{\frac{1 + \alpha^2}{1 + \alpha_c^2}} + \frac{\alpha}{(1 + \sigma^2/\omega^2) \sqrt{1 + \alpha_c^2}} \frac{y - \xi}{\omega}.$$

The proof is by direct computation of the posterior distribution. The statement establishes that the ESN and the normal are conjugate families of distributions, an unusual case given that one of the two components is not of exponential class. On setting $\alpha = 0 = \tau$, one recovers a well-known fact for normal variables. In all cases, α_c is smaller in absolute value than α; in other words, the slant parameter shrinks towards 0.

If one combines Proposition 2.10 with Proposition 2.9, this provides the ingredients for constructing a Kalman-type filter connected to the dynamic linear model:

$$W_t = \rho W_{t-1} + \varepsilon_t,$$

$$Y_t = W_t + \eta_t, \qquad t = 1, 2, \dots$$

where $\{\varepsilon_t\}$ is white noise $N(0, \sigma_\varepsilon^2)$ and $\{\eta_t\}$ is white noise $N(0, \sigma_\eta^2)$. If *a priori* W_0 has a normal distribution, then all subsequent predictive and posterior distributions are still normal, following the classical Kalman filter. If the prior distribution of W_0 is instead taken to be of ESN type, then so is the distribution of W_1, because of Proposition 2.9. Once $Y_1 = y_1$ has been observed, the posterior distribution of W_1 is of ESN type with parameters given by Proposition 2.10. For $t = 2, 3, \ldots$, these features replicate themselves, and the process can be continued. Notice, however, that at each updating stage, the slant parameter shrinks towards 0, and eventually the filter approaches the classical behaviour for normal variates. See the end of §8.2.3 for another form of Kalman filter which avoids this fading phenomenon.

2.3 Historical and bibliographic notes

A discussion on the origins of the SN and ESN families must separate at least two logical perspectives.

The material so far in this chapter has highlighted several and strong connections with the normal distribution, and it is to be expected that the same formal results have been obtained elaborating on normal variables. The following paragraphs summarize work related to the three generating mechanisms described in § 2.1.3 and § 2.2.2. In these contributions, the logical perspective is given by the elaboration of some property of the normal distribution, typically in connection to some motivating problem, but not the construction of a probability distribution more flexible than the normal one, to be used outside the originating problem. It is not surprising that these results have not been developed into an exploration of extensions of the normal family, since this was not the target of the authors.

An alternative view, which aims explicitly to construct supersets of the normal family, has been developed by a more recent stream of literature connected to the framework described in §1.2. In a number of cases, it happened that formal results obtained within this approach turned out in retrospect to coincide with formal results derived earlier under a radically different perspective.

Last in our sequence, but chronologically first, we shall recall some very early work, closely related to our formulation, which could have evolved into a major development if fate had not intervened.

2.3.1 Various origins and different targets

Conditional inspection and selective sampling

Motivated by a practical problem in educational testing, Birnbaum (1950) considered a problem whose essential aspects are as follows. Denote by X_0 the score obtained by a given subject in an attitudinal or educational test, where possibly X_0 is obtained as a linear combination of several such tests, and denote by X_1 the score obtained by the same subject in an admission examination. Assume that, after suitable scaling, (X_0, X_1) is distributed as a bivariate normal random variable with unit marginals and correlation ρ. Since individuals are examined in subsequent tests conditionally on the fact that their admission score exceeds a certain threshold τ', the construction is effectively the same as (2.41) with $\tau = -\tau'$, and the resulting distribution is of type ESN. Besides the density function, Birnbaum (1950) derived some expressions related to moments. This scheme is in turn connected to the question of selective sampling, to be discussed in § 3.4.1.

Recording the largest or the smallest value

Roberts (1966) was concerned with another applied problem, related to observations on twins. Denote by X_0 and X_1 the value taken by a certain variable on a couple of twins, and let $Z_1 = \min\{X_0, X_1\}$ be the quantity of interest. For instance, since twins live in close contact, the occurrence of a cold in one of them very often leads to a cold in the other, but for practical reasons only the age at which the first one gets the cold is recorded. Under the assumption of bivariate normality of (X_0, X_1) with standardized marginal and correlation ρ, Roberts (1966) obtains the distribution of Z_1 following the development leading to (2.16). In addition, he obtains the chi-square property of Proposition 2.1(e) and expressions for the moments of Z_1, including the recurrence (2.63). See Complement 2.3 for a more general form of connection with order statistics of bivariate normal variates.

Additive representation

Weinstein (1964) initiated a discussion in *Technometrics* about the cumulative distribution function of the sum of two independent normal variables, U_0 and U_1, say, when U_1 is truncated by constraining it to exceed a certain threshold. The ensuing discussion, summarized by Nelson (1964), leads to an expression for computing the required probability, which is in essence the distribution function of (2.39).

     Although expressed in a quite different form, a closely related construction has been considered by O'Hagan and Leonard (1976), in a Bayesian context. Denote by θ the mean value of a normal population for which prior considerations suggest that $\theta \geq 0$ but we are not completely confident of this inequality. This uncertainty is handled by a two-stage construction of the prior distribution for θ, assuming that $\theta|\mu \sim N(\mu, \sigma^2)$ and that μ has distribution of type $N(\mu_0, \sigma_0^2)$ truncated below 0. The resulting distribution for θ corresponds again to the sum of a normal and a truncated normal variable.

If the threshold value of the variable U_1 coincides with $\mathbb{E}\{U_1\}$, the above-discussed sum is equivalent to the form $a\,U_0 + b\,|U_1|$, for some real values a and b, and $|U_1| \sim \chi_1$. There is no loss of generality in considering normalized coefficients such that $a^2 + b^2 = 1$, as in (2.14). This special case is directly related to the econometric literature on stochastic frontier analysis, as explained in more detail in § 3.4.2.

A further related case is the threshold autoregressive process studied by Anděl *et al.* (1984) satisfying a relationship essentially of type

$$Z_t = \delta\,|Z_{t-1}| + \sqrt{1 - \delta^2}\,\varepsilon_t \qquad (t = \ldots, -1, 0, 1, \ldots) \qquad (2.50)$$

where the $\{\varepsilon_t\}$'s form a sequence of independent variables $N(0, 1)$. The integral equation for the stationary distribution of the process $\{Z_t\}$ has a solution of type (2.1). For computing the moments, they present an argument which in essence is the one leading to (2.32).

Extending the normal class of distributions

The account presented in the first part of this chapter is based on work connected to the framework of §1.2, and explicitly motivated by the idea of building an extension of the normal class of distributions, at variance with early occurrences described above.

Specifically, §2.1 is largely based on results of Azzalini (1985), with additional material given by the authors indicated next. The additive representation (2.14) has been presented independently by Azzalini (1986) and by Henze (1986); the latter author has also given an expression of the odd moments equivalent to (2.62). The representation via maxima or minima has been presented by Loperfido (2002). Proposition 2.5 has been given by Chiogna (1998), who has also given Proposition 2.3, extending its basic version by Azzalini (1985). The exposition of the ESN distribution in §2.2 is based on Azzalini (1985, Section 3.3) and on Henze (1986), who has given representation (2.43). Additional work has been done by Arnold *et al.* (1993), especially on the statistics side. Some results, such as expressions (2.35) and (2.48) for the distribution function, had appeared earlier in the multivariate context of Chapter 5.

2.3.2 A pioneer

A key idea which forms the basis of this book appeared in some very early work, at the beginning of the 20th century. We are referring to the communication presented by Fernando de Helguero at the Fourth International Congress of Mathematicians held in Rome on 6–11 April 1908. His views were very innovative, and deserve to be presented in some detail. What follows is an excerpt from his written contribution (de Helguero, 1909a), which appeared posthumously because the author died prematurely at the age of 28, in the catastrophic earthquake which hit the town of Messina on 28 December 1908. Another posthumous publication is de Helguero (1909b), which complements the proceedings paper.

The tragic end of de Helguero's life prevented the development of his innovative ideas, and the whole formulation passed unnoticed for the rest of the 20th century. It appears that his contribution has re-emerged only in the discussion of Azzalini (2005), thanks to a personal communication of D. M. Cifarelli.

Sulla rappresentazione analitica delle curve abnormali

Il compito della statistica nelle sue varie applicazioni alle scienze economiche e biologiche non consiste solo nel determinare la legge di dipendenza dei diversi valori ed esprimerla con pochi numeri, ma anche nel fornire un aiuto allo studioso che vuole cercare le cause della variazione e le loro modificazioni. [...]

Invece le curve teoriche studiate dal PEARSON e dall'EDGEWORTH per la perequazione delle statistiche abnormali in materiale omogeneo, mentre dànno con molta approssimazione la legge di variazione (meglio della curva normale perché ne sono delle generalizzazioni), a mio avviso sono difettose in quanto si limitano a dirci che le cause infinitesime elementari della variazione sono *interdipendenti*. Nulla ci fanno sapere sulla legge di dipendenza, quasi nulla

On the analytical representation of abnormal curves

The duty of statistics in its various applications to economics and to biology does not consist only in identifying the law of dependence of the different values and in expressing it with a few numbers, but also in providing some help to the scholar who wants to search the causes of the variation and their modifications. [...]

On the contrary, the theoretical curves studied by PEARSON and by EDGEWORTH for the regularization of abnormal statistics from homogeneous material, while they give with much approximation the law of variation (better than the normal curve because they are generalizations of that one), are defective in my view because they only limit themselves to tell us that the infinitesimal elementary causes of variation are *interdependent*. They tell us nothing on the law of dependence, sulla relazione colla curva normale che pure deve essere considerata fondamentale.

Io penso che miglior aiuto per lo studioso potrebbero essere delle equazioni che supponessero una perturbazione della variabilità normale per opera di cause esterne.

nearly nothing on the connection with the normal curve which still must be considered fundamental.

I think that a better help for the scholar could come from some equations which supposed a perturbation of the normal variability produced by some external causes.

The formulation is distinctly one step ahead of the mainstream approach to data fitting of those years: probability distributions must not simply be devised to provide a numerical fit to observed frequencies, but they must also help to understand how the non-normal distribution has been generated. This goal can be achieved by a formulation which relates perturbation of normality to the effect of some external mechanism.

Of the many hypotheses which can be made on the source of perturbation of the normal distribution, two are discussed by de Helguero: the first form is a mixture of two populations, as it would be called in current terminology, and the second form is via a selection mechanism. The latter form is the one of concern to us; the opening passage of the pertaining section is as follows.

II. Curve perturbate per selezione

Supponiamo che sopra una popolazione distribuita colla legge normale

$$\frac{y_1}{\sigma\sqrt{2\pi}}\, e^{-\frac{1}{2}\left(\frac{x-b}{\sigma}\right)^2}$$

agisca una selezione sfavorevole alle classi più basse (o alle più elevate) tale che per ogni classe y vengano eliminati $y\,\varphi(x)$ individui, dicendo $\varphi(x)$ la probabilità che ha ogni individuo di essere colpito. Noi supponiamo $\varphi(x)$ funzione di x; poiché essa rappresenta una probabilità essa dovrà essere $0 < \varphi(x) < 1$. Per ogni classe rimarranno allora $y - y\varphi(x)$ individui cioè $y(1 - \varphi(x))$ individui.

L'ipotesi più semplice che possiamo fare in $\varphi(x)$ è che sia funzione lineare di x.

$$\varphi(x) = A(x - b) + B\,.$$

Essa acquista il valore zero per $x_0 =$

II. Curves perturbed by selection

Suppose that over a population distributed according to the normal law

$$\frac{y_1}{\sigma\sqrt{2\pi}}\, e^{-\frac{1}{2}\left(\frac{x-b}{\sigma}\right)^2}$$

operates a selection unfavourable to the lower classes (or to the higher ones) such that $y\,\varphi(x)$ individuals are eliminated for any class y, denoting by $\varphi(x)$ the probability that each individual has of being hit. We suppose that $\varphi(x)$ is a function of x; since it represents a probability it must be that $0 < \varphi(x) < 1$. For each class there will then remain $y - y\varphi(x)$ that is $y(1 - \varphi(x))$ individuals.

The simplest hypothesis we can make on $\varphi(x)$ is the one of a linear function of x.

$$\varphi(x) = A(x - b) + B\,.$$

This takes on the value zero at $x_0 =$

$b - \frac{B}{A}$ e il valore 1 per $x_1 = b + \frac{1-B}{A} = x_0 + \frac{1}{A}$ che dovranno perciò cadere fuori del campo di variazione. Sostituendo e ponendo

$$y_0 = y_1(1 - B), \qquad \alpha = -\sigma \frac{A}{1 - B},$$

si ha l'equazione

$$y = \frac{y_0}{\sigma \sqrt{2\pi}} \left(1 + \frac{\alpha(x - b)}{\sigma}\right) e^{-\frac{1}{2}\left(\frac{x-b}{\sigma}\right)^2}.$$

$b - \frac{B}{A}$ and the value 1 at $x_1 = b + \frac{1-B}{A} = x_0 + \frac{1}{A}$, which must therefore lie outside the range of variation. On substituting and setting

$$y_0 = y_1(1 - B), \qquad \alpha = -\sigma \frac{A}{1 - B},$$

one gets the equation

$$y = \frac{y_0}{\sigma \sqrt{2\pi}} \left(1 + \frac{\alpha(x - b)}{\sigma}\right) e^{-\frac{1}{2}\left(\frac{x-b}{\sigma}\right)^2}.$$

As stated in the text, the specific choice of the selection mechanism which operates on a normal density is a very simple one, that is linear (but elsewhere in the text he mentions the possibility of using a different function). The construction is therefore similar to (2.39) with $\Phi(\cdot)$ replaced by the distribution function of a uniform variate over an interval which includes the centre b of the original normal distribution.

In the second part of both papers, de Helguero introduces an additional variant, so that the above expression of the density applies to the half-line, not to a bounded interval. In this sense he is diverging somewhat from the originally planned route summarized above. In spite of this fact, it remains true that he has laid down the essential components of a formulation which extends the normal family of distribution through a selective sampling mechanism of the same type as described earlier in this chapter. In this sense Fernando de Helguero can be considered the precursor of the stream of literature discussed in this work.

For a more detailed discussion of de Helguero's formulation, see Azzalini and Regoli (2012b).

2.4 Some generalizations of the skew-normal family

2.4.1 Preliminary remarks

Start from a larger setting than the title of this section indicates, and consider the distributions of type (1.2) on p. 3 obtained by perturbation of the N(0, 1) distribution. Among the set of densities produced by modulation of a normal base, a fairly natural direction to take is

$$2\,\varphi(x)\,G_0(\alpha\,x)\,, \tag{2.51}$$

replacing Φ with some other symmetric distribution function G_0. A bunch of parametric classes of distributions can then readily be built, taking G_0 equal to the logistic or the Cauchy or the Student's t or the Laplace distribution, and so on.

Some numerical and graphical exploration indicates, however, that the outcome of this process does not appear to produce a set of densities much different from the SN class. More specifically, if we consider two sets of densities formed by the SN densities (2.1) and (2.51) where G_0 is fixed, then for any given choice of α in the SN density there is a suitable choice of λ of the second density which can make the two distributions very similar.

This question has been examined in detail by Umbach (2007); the following summary is somewhat less general than his formulation but it preserves the key features. A sensible criterion for evaluating the dissimilarity between a member of the SN class and a distribution of type (2.51) is

$$\max_x |\Phi(x; \alpha) - \Phi_G(x; \lambda)|,$$

where $\Phi_G(x; \lambda)$ denotes the integral function of (2.51) when the slant parameter is λ. It can be shown that the above maximal difference has a stationary point at $x = 0$; in regular cases, this will be the point of maximal difference. We then consider

$$d(\Phi, \Phi_G) = \max_{\alpha > 0} |\Phi(0; \alpha) - \Phi_G(0; \lambda(\alpha))|,$$

where $\lambda(\alpha)$ denotes the value of λ producing the minimal dissimilarity for the given choice of α; only positive α's need to be considered because of the reflection property of (2.51).

An explicit solution of this optimization problem is not feasible, and one must resort to computational methods for each chosen G_0. Because of the close similarity of the standard normal and the logistic distribution, suitably scaled, it is not surprising that the above measure of dissimilarity $d(\Phi, \Phi_G)$ is very small for G_0 logistic, only 0.00414, which occurs for $\alpha = 1.77$, $\lambda = 3.11$; the densities are plotted in the left panel of Figure 2.6. A less predictable case occurs when G_0 is set equal to the Laplace distribution whose shape is distinctly different from the normal one, but still $d(\Phi, \Phi_G)$ is small, only 0.01287, achieved when $\alpha = 1.47$ and $\lambda = 1.93$; the corresponding densities are plotted in the middle panel of Figure 2.6. A case producing a larger dissimilarity is obtained with G_0 equal to the Cauchy distribution function, where $d(\Phi, \Phi_G)$ takes the value 0.03869, achieved for $\alpha = 2.60$, $\lambda = 11.34$; the densities are displayed in the right panel of Figure 2.6.

All the above examples exhibit a dissimilarity from the shape of the skew-normal distribution which ranges from small to negligible. Preference for one or another family of distributions of type (2.51) is then essentially a

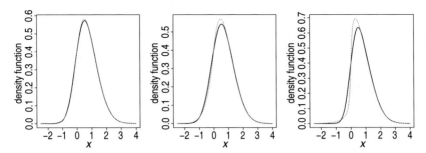

Figure 2.6 Densities of perturbed normal type with linear $w(x)$ having maximal discrepancy from the skew-normal class, when the perturbation cumulative function is a logistic (left), a Laplace (centre) and a Cauchy (right) distribution function.

matter of mathematical tractability. Given its convenient mathematical features, it is reasonable to keep the skew-normal class as the preferred choice.

If we want to obtain a more appreciable change of behaviour, we are then left with the alternative option represented by

$$2\,\varphi(x)\,G_0\{w(x)\},\qquad(2.52)$$

for some non-linear odd function $w(x)$. We shall now present two formulations of this type.

2.4.2 Skew-generalized normal distribution

Arellano-Valle *et al.* (2004) have studied the distribution with density

$$f(x;\alpha_1,\alpha_2) = 2\,\varphi(x)\,\Phi\left(\frac{\alpha_1 x}{\sqrt{1+\alpha_2\,x^2}}\right),\qquad -\infty < x < \infty,\qquad(2.53)$$

where α_1 and α_2 are shape parameters, with $\alpha_2 \geq 0$, and have called it the *skew-generalized normal* (SGN) distribution. Clearly, the case $\alpha_2 = 0$ corresponds to the distribution $SN(0, 1, \alpha_1)$. If $\alpha_1 = 0$, we obtain the $N(0, 1)$ density, irrespective of the value of α_2. Odd moments are only available in an implicit form. Some formal properties of these moments and additional results are given in the above-quoted paper.

Figure 2.7 displays a few examples of SGN densities. In the left panel, $\alpha_1 = 2$ is kept constant with three values of α_2; the right panel is similar with $\alpha_1 = 5$. It is visible that the extra parameter α_2 can be used to regulate the shorter tail of the distribution.

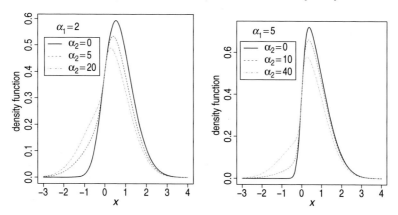

Figure 2.7 SGN density functions having $\alpha_1 = 2$, $\alpha_2 = 0, 5, 20$ (left panel) and $\alpha_1 = 5$, $\alpha_2 = 0, 10, 40$ (right panel).

An interesting property of this distribution is that a variable Z with density (2.53) can be represented as a *shape mixture* of SN variates with suitable distribution of the slant parameter. Specifically, if $(Z|W = w) \sim$ SN$(0, 1, w)$ and $W \sim$ N(α_1, α_2), the marginal density of Z is

$$f(x) = \frac{2}{\sqrt{\alpha_2}} \int_{-\infty}^{\infty} \varphi(x)\, \Phi(x\,w)\, \varphi\left(\frac{w - \alpha_1}{\sqrt{\alpha_2}}\right)\, dw$$

$$= 2 \int_{-\infty}^{\infty} \varphi(x)\, \Phi\left(\sqrt{\alpha_2}\,x\,u + \alpha_1\,x\right)\, du$$

$$= 2\,\varphi(x)\, \mathbb{E}\left\{\sqrt{\alpha_2}\,x\,U + \alpha_1\,x\right\},$$

where $U \sim$ N$(0, 1)$ and, on using Lemma 2.2, $f(x)$ is seen to coincide with (2.53). This representation provides a simple mechanism for sampling data from a SGN distribution. Arellano-Valle *et al.* (2009) have considered similar shape mixtures in the multivariate setting of Chapter 5, and shown how they can be employed for Bayesian inference on the slant parameter.

A reversal of the role of conditioning variable between W and Z exhibits another interesting connection, namely direct calculation gives

$$(W|Z = z) \sim \text{SN}\left(\alpha_1, \alpha_2, \sqrt{\alpha_2}\,z, \frac{\alpha_1\,z}{\sqrt{1 + \alpha_2\,z^2}}\right).$$

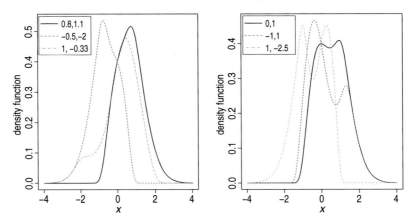

Figure 2.8 Some FGSN density functions with $K = 3$ for a few choices of (α_1, α_3), with unimodal densities in the left panel and bimodal densities in the right panel.

2.4.3 FGSN distribution

A mathematically simple, yet very flexible, choice for $w(x)$ in (2.52) is a polynomial of the form

$$w_K(x) = \alpha_1 x + \alpha_3 x^3 + \cdots + \alpha_K x^K, \tag{2.54}$$

where only odd-order coefficients are included. The distributions identified by (2.52)–(2.54) have been studied by Ma and Genton (2004), who have denoted them *flexible generalized skew-normal* (FGSN) distributions. This is a subset of the more general construction studied by Ma and Genton to be discussed in § 7.2.1.

Since $K = 1$ leads us back to the SN family, the first case to consider is the two-parameter distribution with $K = 3$. This already generates a variety of radically different shapes, as demonstrated by the curves plotted in Figure 2.8, for some pairs (α_1, α_3). It is visible that some choices of (α_1, α_3) produce a unimodal density, while others lead to a bimodal density. Ma and Genton prove that not more than two modes can occur when $K = 3$. Unfortunately, no simple rule is available to tell us whether a given pair (α_1, α_3) corresponds to one mode or two modes.

2.5 Complements

Complement 2.1 (Characteristic function) To compute the characteristic function $Z \sim \mathrm{SN}(0, 1, \alpha)$, Arnold and Lin (2004) make use of the general result that, if the moment generating function $M_X(t)$ of a variable X exists in a neighbourhood of $t = 0$, the characteristic function of X can be computed as $\Psi_X(t) = M_X(i\,t)$. Then, from (2.5), write the moment generating function of Z as

$$M_Z(t) = 2 \, \exp\left(\tfrac{1}{2}t^2\right) \left(\frac{1}{2} + \int_0^{\delta t} \varphi(x)\,\mathrm{d}x\right), \qquad t \in \mathbb{R}.$$

Next, consider the line segment γ linking 0 and δti, namely, γ consists of points $z = xi$, where x takes values from 0 to δt. Then the characteristic function is given by

$$\Psi_Z(t) = M_X(i\,t) = 2 \, \exp\left(-\tfrac{1}{2}t^2\right) \left(\frac{1}{2} + \int_\gamma \varphi(z)\,\mathrm{d}z\right)$$

$$= \exp\left(-\tfrac{1}{2}t^2\right)\left(1 + 2i \int_0^{\delta t} \frac{1}{\sqrt{2\pi}} \exp(x^2/2)\,\mathrm{d}x\right)$$

$$= \exp\left(-\tfrac{1}{2}t^2\right)\{1 + i\,\mathcal{T}(\delta t)\},$$

where

$$\mathcal{T}(x) = b \int_0^x \exp(u^2/2)\,\mathrm{d}u \quad \text{and} \quad \mathcal{T}(-x) = -\mathcal{T}(x), \quad \text{for } x \geq 0.$$

The same result had been obtained earlier by Pewsey (2000b; 2003) by direct evaluation of $\Psi_Z(t) = \mathbb{E}\{\cos(t\,Z)\} + i\,\mathbb{E}\{\sin(t\,Z)\}$. Additional forms of computation have been considered by Kim and Genton (2011), who also obtain the characteristic function of some other distributions presented in later chapters.

Complement 2.2 (Incomplete SN moments)   Chiogna (1998) tackles computation of the incomplete moments of the skew-normal distribution truncated above h,

$$\mu_{z,r}(h) = \int_{-\infty}^h x^r\,\varphi(x;\alpha)\,\mathrm{d}x \qquad (r = 0, 1, 2, \ldots), \tag{2.55}$$

starting from the derivative

$$\frac{\mathrm{d}}{\mathrm{d}x}\varphi(x;\alpha) = -x\,\varphi(x;\alpha) + b\,\alpha\,\varphi(x\sqrt{1+\alpha^2})$$

whose integration over $(-\infty, h)$ gives the incomplete mean value

$$\mu_{z,1}(h) = -\varphi(h;\alpha) + \mu_z\,\Phi(h\sqrt{1+\alpha^2}). \tag{2.56}$$

Integration of $\mu_{z,r}(h)$ by parts, taking into account (2.56), gives the recurrence relationship

$$\mu_{z,r}(h) = -x^{r-1}\,\varphi(x;\alpha) + \mu_z(1 - \mu_z)^{(r-1)/2}\,\mu_{N,r-1}(h\sqrt{1 + \alpha^2})$$
$$+(r-1)\mu_{z,r-2}(h) \tag{2.57}$$

for $r = 2, 3, \ldots$, where $\mu_{N,r}(h)$ denotes the incomplete moment of the ordinary normal distribution, studied by Elandt (1961).

The incomplete moments are directly related to the moments of a truncated skew-normal distribution. The similar case of a doubly truncated distribution, that is with support on a bounded interval, has been examined by Flecher *et al.* (2010).

Complement 2.3 (Connection with normal order statistics)   Representation (2.16) via a maximum or a minimum is a special case of the linear combination of normal order statistics. Nagaraja (1982) has studied the distribution $Z = a_1 Z_{(1)} + a_2 Z_{(2)}$, where $Z_{(1)} \leq Z_{(2)}$ denote the ordered components of a bivariate normal variate with standardized marginals and correlation ρ and a_1, a_2 are constants. Rephrased in our notation, his result is that $Z \sim SN(0, \omega^2, \eta)$ where

$$\omega^2 = a_1^2 + 2\rho a_1 a_2 + a_2^2, \qquad \eta = \sqrt{\frac{1-\rho}{1+\rho}}\,\frac{a_2 - a_1}{\omega(a_1 + a_2)}$$

if $a_1, a_2 \neq 0$ and $a_1^{-1} + a_2^{-1} > 0$. A similar expression holds for $a_1^{-1} + a_2^{-1} < 0$.

Complement 2.4 (SN tail behaviour)   To examine the tail behaviour of the $SN(0, 1, \alpha)$ distribution function, start by noticing that

$$1 - \Phi(x;\alpha) \leq \int_x^\infty 2\varphi(z)\,dz = 2\{1 - \Phi(x)\},$$

for $x \in \mathbb{R}$ and $\alpha \in \mathbb{R}$. Moreover, under the constraint $\alpha < 0$, we can write

$$1 - \Phi(x;\alpha) < 2\Phi(\alpha x)\int_x^\infty \varphi(z)\,dz = 2\,\Phi(\alpha x)\{1 - \Phi(x)\} < 1 - \Phi(x)$$

when $x > 0$. Combining this fact with (B.3) on p. 232, one arrives at

$$1 - \Phi(x;\alpha) < \begin{cases} x^{-1}\,\varphi(x) & \text{if } \alpha < 0, \\ 2\,x^{-1}\,\varphi(x) & \text{if } \alpha > 0. \end{cases} \tag{2.58}$$

These inequalities already provide upper bounds for the upper tail probabilities of the SN distribution for negative and for positive α, respectively, or equivalently bounds of the two tails at any given α, because of Proposition 2.7(a).

With further algebraic work, we can obtain that

$$q(x, \alpha)\, r(x, \alpha) < 1 - \Phi(x; \alpha) < q(x, \alpha), \qquad \text{if } \alpha < 0,\ x > 0 \qquad (2.59)$$

and

$$2\, \frac{\varphi(x)}{x}\left(1 - \frac{1}{x^2}\right) - q(x, \alpha) < 1 - \Phi(x; \alpha) < 2\, \frac{\varphi(x)}{x} - q(x, \alpha)\, r(x, \alpha),$$
$$\text{if } \alpha > 0,\ x > 0, \qquad (2.60)$$

where $q(x, \alpha)$ is given by (2.38) and

$$r(x) = 1 - \frac{1 + 3\alpha^2}{x^2\, \alpha^2 (1 + \alpha^2)}.$$

These inequalities show that the right tail decreases at the same rate as the normal distribution tail, when $\alpha > 0$, while the left tail has a faster rate of convergence to 0. Proposition 2.8 follows as an immediate corollary of (2.59) and (2.60). For details of the above development and for improved bounds, see Capitanio (2010).

The results of Proposition 2.8 allow us to prove quite easily that the distribution function $\Phi(x; \alpha)$ belongs to the domain of attraction of the Gumbel distribution, similarly to the normal; see Problem 2.10.

Chang and Genton (2007) arrive at the same result via a different route; see their Proposition 3.1. They also obtain the general result concerning the tail behaviour of the FGSN distribution defined in § 2.4.3. See also Padoan (2011, p. 979) for further details.

Complement 2.5 (Log-skew-normal distribution)   A mention is due of the distribution arising from exponentiation of $Y \sim \mathrm{SN}(\xi, \omega^2, \alpha)$, even if technically the outcome does not fall within the formulation (1.2). By analogy with the log-normal distribution, we shall say $X = \exp(Y)$ is a log-skew-normal variate. The density function of X is

$$f_X(x) = \frac{1}{x\,\omega}\, \varphi\left(\frac{(\log x) - \xi}{\omega}; \alpha\right), \qquad x \in (0, \infty). \qquad (2.61)$$

The kth moment of X is readily obtained by evaluating (2.5) at $t = k$; this leads to

$$\mathbb{E}\{X\} = 2\, \exp(\xi + \tfrac{1}{2}\omega^2)\, \Phi(\delta\omega),$$
$$\mathrm{var}\{X\} = 2\, \exp(2\,\xi)\left[\exp(2\omega^2)\, \Phi(2\delta\omega) - 2\, \exp(\omega^2)\, \Phi(\delta\omega)^2\right].$$

Numerical work fitting (2.61) to the distribution of family income data has been done by Azzalini *et al.* (2003); see also Chai and Bailey (2008). Similarly to the log-normal distribution, (2.61) is moment-indeterminate (Lin and Stoyanov, 2009).

Problems

2.1 If $Z|\alpha \sim \text{SN}(0, 1, \alpha)$ and α is a continuous random variable whose density function $g(\alpha)$ is symmetric about 0, then use Proposition 1.1 to conclude that the unconditional distribution of Z is N(0, 1), irrespective of $g(\alpha)$.

2.2 As remarked in § 2.1.2, the convolution of two SN distributions is not of SN type. This points against the conjecture that the distribution is infinitely divisible: confirm this fact (Domínguez-Molina and Rocha-Arteaga, 2007; Kozubowski and Nolan, 2008).

2.3 For $\zeta_2(x)$ defined in (2.20), prove that $-1 < \zeta_2(x) < 0$ for all real x.

2.4 Other results similar to Proposition 2.5 are as follows. If $Z \sim \text{SN}(0, 1, \alpha)$, then prove that

$$\mathbb{E}\{\Phi(hZ; \beta)\} = \frac{1}{2} - \frac{1}{\pi} \arctan \frac{r\beta - hq\alpha}{rq + hq\beta},$$

$$\mathbb{E}\{\Phi(hZ)^2\} = \frac{1}{4} + \frac{1}{\pi} \arctan \frac{h\alpha}{r} + \frac{1}{2\pi} \arctan \frac{h^2}{\sqrt{1 + 2h^2}},$$

where $q = \sqrt{1 + h^2(1 + \beta^2)}$ and $r = \sqrt{1 + h^2 + \alpha^2}$ (Chiogna, 1998).

2.5 Show that the odd moments of $Z \sim \text{SN}(0, 1, \alpha)$ can be written as

$$\mathbb{E}\{Z^{2m+1}\} = b\delta(1 - \delta^2)^m \frac{(2m + 1)!}{2^m} \sum_{j=0}^{m} \frac{j!}{(2j + 1)!(m - j)!} \left(\frac{4\delta^2}{1 - \delta^2}\right)^j$$

$$= \mu_{z,2m+1}, \qquad\qquad\qquad (2.62)$$

say, for $m = 0, 1, \ldots$ (Henze, 1986).

2.6 Show that the odd moments given by (2.62) satisfy the recursive relationship

$$\mu_{z,2m+1} = 2m\,\mu_{z,2m-1} + b\frac{(2m)!}{2^m\,m!}\delta(1 - \delta^2)^m, \qquad m = 1, 2, \ldots \quad (2.63)$$

which we can start from $\mu_{z,1}$ given by (2.26) (Roberts, 1966; Martínez *et al.*, 2008).

2.7 Use (2.56) to show that the absolute mean value $\mathbb{E}\{|Z - t|\}$ from an arbitrary constant t, when $Z \sim \text{SN}(0, 1, \alpha)$, is

$$\mathbb{E}\{|Z - t|\} = \mu_z - t + 2\{\varphi(t; \alpha) + t\,\Phi(t; \alpha) - \mu_z\,\Phi(t\sqrt{1 + \alpha^2})\}.$$

In the special case with $t = \mu_z$, we obtain the mean absolute deviation (from the mean)

$$\mathbb{E}\{|Z - \mu_z|\} = 2\left[\varphi(\mu_z; \alpha) + \mu_z\left\{\Phi(\mu_z; \alpha) - \Phi\left(\mu_z\sqrt{1+\alpha^2}\right)\right\}\right] \approx b\,\sigma_z.$$

The last expression provides a simple but effective approximation to the exact value, and is exact for $\alpha = 0$ (Azzalini *et al.*, 2010).

2.8 Show that the incomplete mean value of the ESN distribution truncated above h is

$$\delta\,\zeta_1(\tau)\,\Phi(h\sqrt{1+\alpha^2} + \tau\alpha) - \varphi(h; \alpha, \tau), \qquad (2.64)$$

which reduces to (2.56) when $\tau = 0$. [*Hints*: Use the argument leading to (2.56); alternatively, use (B.22) on p. 235.]

2.9 From the relationships (B.20)–(B.27) one can derive various interesting facts. For instance, the alternative form of the ESN distribution function (2.48) can be obtained by an elementary manipulation of (B.21), on setting $a = \tau\sqrt{1+\alpha^2}$, $b = \alpha$. Using (B.25) show that

$$\Phi(x; \alpha, \tau)\,\Phi(\tau) = \Phi(\tau; \alpha, x)\,\Phi(x),$$

which at $x = 0$ gives

$$\Phi(0; \alpha, \tau) = \frac{\Phi(\tau; \alpha)}{2\,\Phi(\tau)} = \frac{1}{2} - \frac{T(\tau, \alpha)}{\Phi(\tau)};$$

see also Canale (2011). From (B.28) show that

$$\Phi(x; \alpha, x)\,\Phi(x) = \Phi\left(x; \alpha + \sqrt{1+\alpha^2}\right).$$

2.10 Theorem 1.6.1 of Leadbetter *et al.* (1983) states that a distribution function F belongs to the domain of attraction of the Gumbel distribution if

$$\lim_{x\to\infty} \frac{[1 - F(x)]\,f'(x)}{f(x)^2} = -1,$$

where $f(x) = F'(x)$. Use this fact and the tail approximation given in Proposition 2.8 to prove that the SN distribution function $\Phi(x; \alpha)$ belongs to the domain of attraction of the Gumbel distribution, similarly to the normal.

2.11 Starting from an arbitrary density function $p_0(x)$ with moment generating function $M_0(\cdot)$, exponential tilting refers to the exponential family of densities defined by

$$p(x; \theta) = \exp(\theta x)\,p_0(x)/M_0(\theta)$$

indexed by the parameter θ (Efron, 1981). Show that if we set p_0

equal to the SN density $\varphi(x; \alpha)$, the corresponding exponentially tilted distribution is $\mathrm{SN}(\theta, 1, \alpha, \delta\theta)$, where $\delta = \delta(\alpha)$. Although in general the ESN distribution does not have an exponential family structure, this is the case for this specific subclass, when α is fixed and θ is a free parameter (Dalla Valle, 1998, pp. 82–84).

3

The skew-normal distribution: statistics

The preceding chapter has shown how similarly the skew-normal distribution behaves to the classical normal one from the viewpoint of probability. In this chapter we shall deal with the statistical aspects, and a radically different picture will emerge.

3.1 Likelihood inference

Our primary approach to statistical methodology is via likelihood-based inference. The concepts which we shall make use of and their notation are quite standard; however, for completeness, they are recalled briefly in Appendix C. The only apparently unusual quantity is the deviance function; see (C.12) on p. 239 and (C.16).

3.1.1 The log-likelihood function

If y denotes a value sampled from a random variable $Y \sim SN(\xi, \omega^2, \alpha)$, its contribution to the log-likelihood function is

$$\ell_1(\theta^{\mathrm{DP}}; y) = \text{constant} - \log \omega - \frac{(y - \xi)^2}{2\,\omega^2} + \zeta_0\left(\alpha \frac{y - \xi}{\omega}\right), \qquad (3.1)$$

where $\theta^{\mathrm{DP}} = (\xi, \omega, \alpha)^{\mathsf{T}}$ and $\zeta_0(\cdot)$ is defined by (2.18) on p. 30. The superscript 'DP' stands for *direct parameters*; the motivation for this term will become clear later on. If $z = (y - \xi)/\omega$ and $\zeta_1(\cdot)$ is defined by (2.20), the components of the score vector are

$$\frac{\partial \ell_1}{\partial \xi} = \frac{z}{\omega} - \frac{\alpha}{\omega}\zeta_1(\alpha z),$$

$$\frac{\partial \ell_1}{\partial \omega} = -\frac{1}{\omega} + \frac{z^2}{\omega} - \frac{\alpha}{\omega}\zeta_1(\alpha z)\,z, \qquad (3.2)$$

$$\frac{\partial \ell_1}{\partial \alpha} = \zeta_1(\alpha z)\,z.$$

If a random sample $y_1, \ldots, y_n$ from $Y \sim \mathrm{SN}(\xi, \omega^2, \alpha)$ is available, the log-likelihood $\ell(\theta^{\mathrm{DP}})$ is obtained by summation of n terms of type (3.1) and a corresponding sum of terms (3.2) leads to the likelihood equations

$$
\begin{aligned}
\sum_i z_i - \alpha \sum_i \zeta_1(\alpha z_i) &= 0, \\
\sum_i z_i^2 - \alpha \sum_i z_i \zeta_1(\alpha z_i) &= n, \\
\sum_i z_i \zeta_1(\alpha z_i) &= 0,
\end{aligned}
\tag{3.3}
$$

where $z_i = (y_i - \xi)/\omega$, for $i = 1, \ldots, n$. The presence of the non-linear function ζ_1 prevents explicit solution of these equations, and numerical methods must be employed. Note that, for a point which is a solution of the third equation, the second equation requires that ξ and ω satisfy

$$
\hat{\omega}^2 = \frac{1}{n} \sum_i (y_i - \hat{\xi})^2
\tag{3.4}
$$

which reproduces a well-known fact for normal variates, and effectively removes the second equation in (3.3).

Another simple remark is that, when $\alpha = 0$, (3.1) reduces to the log-likelihood function for the normal distribution, and it is well known that in the normal case the maximum is achieved at $\xi = \bar{y}$, $\omega = s$, where

$$
\bar{y} = \frac{1}{n} \sum_{i=1}^n y_i, \qquad s = \left(\frac{1}{n} \sum_{i=1}^n (y_i - \bar{y})^2 \right)^{1/2}
\tag{3.5}
$$

denote the sample mean and the uncorrected sample standard deviation. Moreover, the point $\theta^{\mathrm{DP}} = (\bar{y}, s, 0)^\top$ is also a solution of the third of equations (3.3), and therefore it is a stationary point of $\ell(\theta^{\mathrm{DP}})$, for *any* sample. The argument is mathematically elementary, but the consequences are non-trivial, as will emerge in the subsequent development.

If $\hat{\xi}(\alpha)$ and $\hat{\omega}(\alpha)$ denote the maximum likelihood estimate (MLE) of ξ and ω, for any fixed value of α, the profile log-likelihood function is

$$
\ell^*(\alpha) = \ell\big(\hat{\theta}^{\mathrm{DP}}(\alpha)\big),
\tag{3.6}
$$

where $\hat{\theta}^{\mathrm{DP}}(\alpha) = (\hat{\xi}(\alpha), \hat{\omega}(\alpha), \alpha)^\top$. A closely related conclusion of the above facts on the efficient scores holds for $\ell^*(\alpha)$, whose derivative is

$$
\frac{d\ell^*(\alpha)}{d\alpha} = \frac{\partial \ell(\theta^{\mathrm{DP}})}{\partial \xi} \frac{d\hat{\xi}(\alpha)}{d\alpha} + \frac{\partial \ell(\theta^{\mathrm{DP}})}{\partial \omega} \frac{d\hat{\omega}(\alpha)}{d\alpha} + \frac{\partial \ell(\theta^{\mathrm{DP}})}{\partial \alpha},
\tag{3.7}
$$

where the partial derivatives are evaluated at $\theta^{\mathrm{DP}} = \hat{\theta}^{\mathrm{DP}}(\alpha)$. From (3.2), it is immediate that these partial derivatives vanish at $\hat{\theta}^{\mathrm{DP}}(0) = (\bar{y}, s, 0)^\top$, and it then follows that $\ell^*(\alpha)$ always has a stationary point at $\alpha = 0$. An obvious implication is that the score test for the null hypothesis $H_0 : \alpha = 0$ is void, at least in its standard form.

Notice that the crucial feature in these peculiar aspects of the log-likelihood function is the proportionality of the first and third component of (3.2) when $\alpha = 0$.

We extend our formulation slightly to include the case of a linear regression setting for the location parameter ξ, which is expressed as a linear combination of a p-dimensional set of covariates x, that is

$$\xi = x^\top \beta, \qquad \beta \in \mathbb{R}^p, \tag{3.8}$$

where β is an unknown parameter vector. In this case, the first expression of the score function (3.2) is replaced by

$$\frac{\partial \ell_1}{\partial \beta} = \left(\frac{z}{\omega} - \frac{\alpha}{\omega} \zeta_1(\alpha z) \right) x. \tag{3.9}$$

In this setting, we assume that n $(n > p)$ independently sampled observations $y = (y_1, \ldots, y_n)^\top$ are available, with associated $n \times p$ design matrix $X = (x_1, \ldots, x_n)^\top$ of rank p. To simplify the treatment, we also assume that 1_n belongs to the space spanned by the columns of X, a condition satisfied in nearly all cases; typically, 1_n is the first column of X. In the regression case, (3.4) must be modified to

$$\hat{\omega}^2 = \frac{1}{n} \sum_i (y_i - \hat{\xi}_i)^2 \tag{3.10}$$

where $\hat{\xi}_i = x_i^\top \hat{\beta}$ for $i = 1, \ldots, n$.

When $\alpha = 0$, summation of n terms of type (3.9) leads to the estimating equations $\sum_i z_i x_i = 0$, that is the normal equations of linear models. Hence a stationary point of the log-likelihood occurs at $\tilde{\theta}^{\text{DP}} = (\tilde{\beta}^\top, s, 0)^\top$, where $\tilde{\beta} = (X^\top X)^{-1} X^\top y$ is the least-squares estimate and in this case s is given by the uncorrected standard deviation of the least-squares residuals, $y - X\tilde{\beta}$.

3.1.2 A numerical illustration

Before entering additional aspects, it is useful to illustrate what we have seen so far with the aid of a numerical example. Guided by the Latin saying *cave nil vino*,[1] we base our illustration on some measurements taken from a set of 178 specimens of Italian wines presented by Forina *et al.* (1986). The data refer to three cultivars of the Piedmont region, namely Barbera,

---

[1] Beware of the lack of wine.

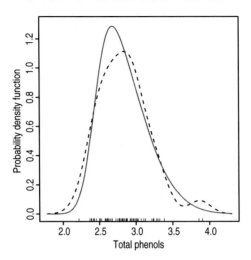

Figure 3.1 Wines data: total phenols content of Barolo. The observations are denoted by ticks on the horizontal axis, the dashed line denotes a non-parametric estimate of the density function, the solid line denotes the SN density selected by maximum likelihood estimation.

Barolo and Grignolino. From these cultivars, 48, 59 and 71 specimens, respectively, have been collected, followed by the extraction of 28 chemical measurements from each specimen.

Figure 3.1 refers to $n = 59$ measurements on total phenols in the Barolo samples. Phenols are important constituents of wine chemistry. The individual measurements are marked by ticks on the abscissa; the dashed line represents a non-parametric density estimate; the solid line corresponds to the skew-normal density selected by maximum likelihood estimation, which will be discussed in more detail shortly.

Inspection of the non-parametric estimate indicates departure from normality of the distribution, largely in the form of presence of moderate but clearly visible skewness. This indication is further supported by the sample coefficient of skewness,

$$\bar{\gamma}_1 = \frac{\sum_i (y_i - \bar{y})^3 / n}{s^3}, \tag{3.11}$$

which in this case equals 0.795. If this value is standardized with its asymptotic standard error under assumption of normality, $\sqrt{6/n}$, we obtain 2.49, which confirms the indication of an asymmetric distribution, although not in an extreme form. A more refined form of standardization could be employed, based on the exact variance of $\bar{\gamma}_1$ under normality; see Cramér (1946, p. 386). With this refinements the standardized value increases slightly to 2.62.

To fit a skew-normal distribution to the data, numerical maximization of (3.1) produces the estimate $\hat{\theta}^{\text{DP}} = (2.44, 0.521, 3.25)^{\top}$. This corresponds to the SN density depicted by the solid line in Figure 3.1, which appears to follow the non-parametric curve reasonably well.

Another, and in a sense more informative, type of graphical diagnostics of adequacy of the fitted distribution can be produced starting from the normalized residuals

$$\hat{z}_i = (y_i - \hat{\xi})/\hat{\omega} \qquad (i = 1, \ldots, n), \tag{3.12}$$

such that $\hat{z}_i^2$ should be sampled from an approximate χ_1^2 distribution, recalling that

$$Z^2 = (Y - \xi)^2/\omega^2 \sim \chi_1^2 \tag{3.13}$$

from Proposition 2.1(e). Therefore, we construct a plot of the points $(q_i, \hat{z}_{(i)}^2)$ where q_i denotes the quantile of level $i/(n+1)$ of the χ_1^2 distribution, and $\hat{z}_{(i)}^2$ is the ith largest $\hat{z}_i^2$, for $i = 1, \ldots, n$. This device is called a QQ-plot since empirical quantiles are plotted versus theoretical quantiles. If the SN assumption holds true, we expect that the points tend to be aligned along the identity line. In other words, we are essentially replicating the same construction leading to the half-normal probability plot, based on the absolute values of the usual standardized residuals

$$r_i = (y_i - \bar{y})/s \qquad (i = 1, \ldots, n). \tag{3.14}$$

The two plots displayed in Figure 3.2 have been constructed in this way, with the left-hand plot based on the $\hat{z}_i$ values and the right-hand plot based on the r_i values. The points of the first plot are more closely aligned along the identity line than those of the second plot, indicating that the skew-normal distribution provides a better fit than the normal one.

It is also instructive to visualize the log-likelihood function. Since there are three components in θ^{DP}, this is not feasible for $\ell(\theta^{\text{DP}})$ directly, and we must consider plots which reduce dimensionality. The usual option is to consider profile log-likelihoods, and corresponding deviances.

The left-hand panel of Figure 3.3 displays the deviance function $D(\alpha)$ associated with the profile log-likelihood (3.6). The function is noticeably non-quadratic: one feature is that the function increases on the right of the minimum at $\alpha = 3.25$ more gently than on the left of $\hat{\alpha}$, but the more

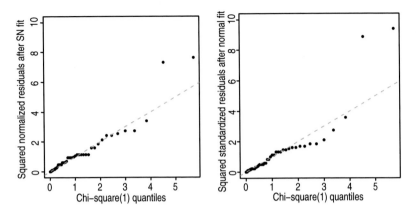

Figure 3.2 Wines data: phenols content of Barolo. QQ-plot of squared residuals under the assumption of skew-normal (left panel) and normal distribution (right panel); the dashed line represents the identity function.

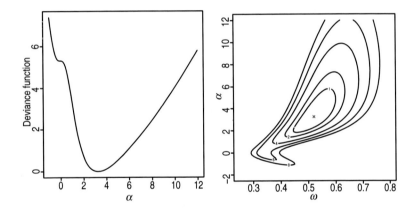

Figure 3.3 Wines data: phenols content of Barolo. Profile deviance function for the parameter α of the SN distribution in the left panel and for (ω, α) in the right panel; the mark $\times$ indicates $(\hat{\omega}, \hat{\alpha})$.

peculiar feature is the stationary point at $\alpha = 0$, as expected from the discussion in § 3.1.1.

Recall that, in a k-parameter regular estimation problem, the set of parameter values having deviance not larger than the pth quantile of the χ_k^2 distribution delimit a confidence region with approximate confidence level p. Correspondingly, the deviance function can be used for hypothesis testing by examining whether a nominated point belongs to the confidence region so obtained. In the present setting, the above-remarked non-regularity of the deviance function affects the construction of confidence sets, which in the present case with $k = 1$ usually correspond to intervals. It is not a workaround to build a confidence interval in the form $\hat{\theta} \pm 2(\text{std.err.})$ since the validity of this method relies anyway on regularity of the log-likelihood, hence of the deviance.

One could remark that the stationary point at $\alpha = 0$ occurs where $D(\alpha)$ is quite large, above 5, hence in a region which does not effectively interfere with our construction of confidence intervals and hypothesis testing, at least for the confidence levels usually considered. It is, however, clear that the problem still exists for other data sets with $\hat{\alpha}$ closer to 0, for which $D(0)$ will not be so large.

The right-hand panel of Figure 3.3 displays the profile deviance function for the parameter pair (ω, α). The regions delimited by the contour lines of $D(\omega, \alpha)$ are not only markedly different from the ideal elliptical shape, typical of linear normal models, but some of them are not even convex regions. Notice that this non-convexity is associated with the crossing of the line $\alpha = 0$.

The overall message emerging from the plots in Figure 3.3 is that the log-likelihood function associated with SN variates is somehow unusual. This is an aspect to be examined more closely in the next sections.

Before entering technical aspects, it is advisable to underline a qualitative effect of working with a parametric family which effectively is regulated by moments up to the third order. The implication is that the traditional rule of thumb by which a sample size is small up to 'about $n = 30$', and then starts to become 'large', while sensible for a normal population or other two-parameter distribution is not really appropriate here. To give an indication of a new threshold is especially difficult, because the value of α also has a role here. Under this *caveat*, numerical experience suggests that 'about $n = 50$' may be a more appropriate guideline in this context.

3.1.3 Fisher information for the direct parameterization

Consider the linear regression setting introduced near the end of § 3.1.1 and assume that a set of n observations $y = (y_1, \ldots, y_n)^\top$ is available, drawn under independent sampling. The contributions from a single observation to the score functions for the direct parameter $\theta^{\mathrm{DP}} = (\beta^\top, \omega, \alpha)^\top$ are given by (3.9) and the last two expressions in (3.2), respectively. Differentiation of these score functions, summed over n observations, leads to

$$-\frac{\partial^2 \ell}{\partial \beta \, \partial \beta^\top} = \omega^{-2} X^\top (I_n + \alpha^2 Z_2) X \,,$$

$$-\frac{\partial^2 \ell}{\partial \beta \, \partial \omega} = \omega^{-2} X^\top (2z - \alpha \zeta_1(\alpha z) + \alpha^2 Z_2 z) \,,$$

$$-\frac{\partial^2 \ell}{\partial \beta \, \partial \alpha} = \omega^{-1} X^\top (\zeta_1(\alpha z) - \alpha Z_2 z), \tag{3.15}$$

$$-\frac{\partial^2 \ell}{\partial \omega^2} = \omega^{-2} \left(-n + 3 \, (1_n^\top z^2) - 2\alpha \zeta_1(\alpha z)^\top z - \alpha^2 \zeta_2(\alpha z)^\top z^2 \right) \,,$$

$$-\frac{\partial^2 \ell}{\partial \omega \, \partial \alpha} = \omega^{-1} \left(\zeta_1(\alpha z)^\top z + \alpha \, \zeta_2(\alpha z)^\top z^2 \right) \,,$$

$$-\frac{\partial^2 \ell}{\partial \alpha^2} = -\zeta_2(\alpha z)^\top z^2 \,,$$

where 1_n denotes the n-dimensional vector of all 1's,

$$z = \omega^{-1}(y - X\beta) \,, \qquad Z_2 = \mathrm{diag}(-\zeta_2(\alpha z)) > 0,$$

and we adopt the convention that z^2 is the vector obtained by squaring each element of z and similarly an expression of type $\zeta_k(\alpha z)$ represents the vector obtained by applying the function $\zeta_k(\cdot)$ to each element of αz.

Evaluation of the mean value of the above second derivatives involves expectations of some non-linear functions of $Z \sim \mathrm{SN}(0, 1, \alpha)$. Some terms are simple to obtain, specifically

$$
\begin{aligned}
\mathbb{E}\{Z^k \, \zeta_1(\alpha Z)\} &= \frac{b}{(1 + \alpha^2)^{(k+1)/2}} \, \mathbb{E}\{U^k\} \\
&= \frac{b}{(1 + \alpha^2)^{(k+1)/2}} \begin{cases} 1 \times 3 \times \cdots \times (k-1) & \text{if } k = 0, 2, 4, \ldots, \\ 0 & \text{if } k = 1, 3, 5, \ldots, \end{cases}
\end{aligned}
$$

where $U \sim N(0, 1)$ and $b = \sqrt{2/\pi}$. Other terms are not so manageable, specifically the quantities

$$a_k = a_k(\alpha) = \mathbb{E}\{Z^k \, \zeta_1(\alpha Z)^2\} \,, \qquad k = 0, 1, \ldots, \tag{3.16}$$

which we need to compute numerically for $k = 0, 1, 2$. With these elements and recalling that $\zeta_2(u) = -\zeta_1(u)\{u + \zeta_1(u)\}$, the expected Fisher information matrix is

$$
\mathcal{I}^{\text{DP}}(\theta^{\text{DP}}) = \begin{pmatrix} \dfrac{1 + \alpha^2 a_0}{\omega^2} X^\top X & \cdot & \cdot \\ \dfrac{1}{\omega^2}\left(\dfrac{b\,\alpha(1 + 2\alpha^2)}{(1+\alpha^2)^{3/2}} + \alpha^2\,a_1\right) 1_n^\top X & n\,\dfrac{2+\alpha^2 a_2}{\omega^2} & \cdot \\ \dfrac{1}{\omega}\left(\dfrac{b}{(1+\alpha^2)^{3/2}} - \alpha\,a_1\right) 1_n^\top X & -n\,\dfrac{\alpha\,a_2}{\omega} & n\,a_2 \end{pmatrix}, \quad (3.17)
$$

where the upper triangle must be completed by symmetry.

Given the peculiar aspects that have emerged in the previous sections when $\alpha = 0$, consider this case more closely. The expected information matrix reduces then to

$$
\mathcal{I}^{\text{DP}}((\beta^\top, \omega, 0)^\top) = \begin{pmatrix} \omega^{-2} X^\top X & 0 & \omega^{-1} b\,1_n^\top X \\ 0 & \omega^{-2}\,2n & 0 \\ \omega^{-1} b\,1_n^\top X & 0 & b^2\,n \end{pmatrix}, \quad (3.18)
$$

whose determinant

$$
\frac{2\,n\,b^2}{\omega^{2(p+1)}} \det(X^\top X)\left[n - 1_n^\top X(X^\top X)^{-1} X^\top 1_n\right]
$$

is 0, having assumed that 1_n belongs to the column space of X.

The cause of this quite uncommon phenomenon is easily seen in the simple sample case, whose efficient scores are given by (3.2). When $\alpha = 0$, the first and third components are proportional to each other, for all possible sample values. This implies rank-deficiency of the variance matrix of the score functions, that is, of the expected Fisher information matrix.

When $\alpha = 0$, singularity of the expected information prevents applications of standard asymptotic theory of MLE. Although this anomalous behaviour is limited to the specific value $\alpha = 0$, the fact is nevertheless unpleasant, given that the point $\alpha = 0$ corresponds to the subset of proper normal distributions. For instance, a natural problem to consider is to test the null hypothesis that $\alpha = 0$, but standard methodology does not apply, given the above singularity.

Singularity of the expected information matrix is matched by its sample counterpart, in the following sense. We have seen at the end of § 3.1.1 that $\tilde{\theta}^{\text{DP}} = (\tilde{\beta}^\top, s, 0)^\top$ is always a solution of the likelihood equations. Evaluation of the second derivative at $\tilde{\theta}^{\text{DP}}$ gives

$$
-\frac{\partial^2 \ell}{\partial \theta^{\text{DP}}\,\partial(\theta^{\text{DP}})^\top}\bigg|_{\theta^{\text{DP}} = \tilde{\theta}^{\text{DP}}} = \begin{pmatrix} s^{-2} X^\top X & 0 & s^{-1} b\,X^\top 1_n \\ 0 & s^{-2}\,2n & 0 \\ bs^{-1}\,1_n^\top X & 0 & nb^2 \end{pmatrix}, \quad (3.19)
$$

which is singular too.

3.1.4 Centred parameterization

To overcome the problem of singularity of the information matrix at 0, we must get some insight into the source of the problem. To ease discussion, consider the simple case where y is a random sample of size n from $Y \sim \mathrm{SN}(\xi, \omega^2, \alpha)$, hence $X = 1_n$. Since all moments of Y exist, sample moments are unbiased consistent estimates of the corresponding population moments as $n \to \infty$, with variance proportional to $1/n$.

We focus on $\gamma_1(Y)$, given by (2.28) on p. 31, because it depends on α only. It is well known that $\bar{\gamma}_1$, the sample coefficient of skewness, converges to the true parameter γ_1 with asymptotic variance $6/n$. Inversion of (2.28) gives

$$\alpha = \frac{R}{\sqrt{2/\pi - (1 - 2/\pi) R^2}}, \qquad R = \frac{\mu_z}{\sigma_z} = \sqrt[3]{\frac{2 \gamma_1}{4 - \pi}}, \qquad (3.20)$$

showing that, in a neighbourhood of the origin, α is approximately proportional to $\sqrt[3]{\gamma_1}$. Since this transformation of γ_1 has unbounded derivative at 0, the corresponding sample value of α computed via (3.20), $\bar{\alpha}$, does not converge to 0 at the usual rate, when the true parameter value is 0. More specifically, since $\bar{\gamma}_1$ is $O_p(n^{-1/2})$, then $\bar{\alpha}$ is $O_p(n^{-1/6})$, by the very definition of order in probability. In addition, since

$$\xi = \mu - b\,\omega\delta(\alpha) \approx \mu - b\omega\alpha$$

near $\alpha = 0$, where $\mu = \mathbb{E}\{Y\}$ is given by (2.22), a similar behaviour holds for the estimate of ξ. In essence, the singularity problem is due to the nature of the functions connecting moments and direct parameters; it is not intrinsic to the SN family.

Strictly speaking, the above argument refers to estimation via the method of moments, but in its essence we can retain it as valid also for MLE, since typically the two methods have the same asymptotic distribution, to the first order of approximation.

Motivated by these remarks, we introduce a reparameterization aimed at removing the singularity problem when $\alpha = 0$. Rewrite Y as

$$Y = \mu + \sigma Z_0, \qquad Z_0 = \frac{Z - \mu_z}{\sigma_z} \sim \mathrm{SN}\left(-\frac{\mu_z}{\sigma_z}, \frac{1}{\sigma_z^2}, \alpha\right),$$

where $\sigma^2 = \mathrm{var}\{Y\}$ is given by (2.23). Consider $\theta^{\mathrm{CP}} = (\mu, \sigma, \gamma_1)^{\top}$ as the new parameter vector with admissible set

$$\mathbb{R} \times \mathbb{R}^+ \times (-\gamma_1^{\max}, \gamma_1^{\max}),$$

where $\gamma_1^{\max}$ is given by (2.31). We shall call the components of θ^{CP} *centred parameters* (CP), because their construction involves, at least notionally, the variable Z_0, centred at 0. By contrast, the θ^{DP} components can be read directly from the expression of the density function, hence the name 'direct parameters'. Explicitly, the mapping from DP to CP is

$$
\begin{aligned}
\mu &= \xi + b\,\omega\frac{\alpha}{\sqrt{1+\alpha^2}} = \xi + b\,\omega\,\delta(\alpha)\,, \\
\sigma &= \omega\left(1 - b^2\,\delta(\alpha)^2\right)^{1/2}, \\
\gamma_1 &= \frac{4-\pi}{2}\frac{b^3\,\delta(\alpha)^3}{[1-b^2\,\delta(\alpha)^2]^{3/2}} = \frac{4-\pi}{2}\frac{b^3\,\alpha^3}{[1+(1-b^2)\alpha^2]^{3/2}}
\end{aligned}
\tag{3.21}
$$

and the inverse mapping is provided by (3.20) and

$$
\omega = \frac{\sigma}{(1-b^2\,\delta(\alpha)^2)^{1/2}}\,, \qquad \xi = \mu - b\,\omega\,\delta(\alpha)\,.
\tag{3.22}
$$

Moreover, since the components of CP are smooth functions of the first three moments, or equivalently three cumulants, we can expect them to lead to a regular asymptotic distribution of the MLE, $\hat{\theta}^{\mathrm{CP}}$. We shall discuss this aspect in more detail in the next section.

Another important advantage of the centred parameterization, relevant not only for the above asymptotic considerations connected to the subset with $\alpha = 0$, is that μ is a far more familiar location parameter, usually with a clearer subject-matter interpretation than ξ. For similar reasons, σ and γ_1 are preferable to ω and α, respectively.

In the regression case, assume for simplicity that 1_n is the first column of X, and denote by β_0 the corresponding parameter. The CP formulation is then extended by setting $\theta^{\mathrm{CP}} = (\beta^{\mathrm{CP}}, \sigma, \gamma_1)$ where, in an obvious notation, $\beta^{\mathrm{CP}} = \beta^{\mathrm{DP}}$ for all components except the first one, such that

$$
\beta_0^{\mathrm{CP}} = \beta_0^{\mathrm{DP}} + \omega\,\mu_z
\tag{3.23}
$$

which matches (2.22) on p. 30.

The expected and observed information matrices for CP can be obtained from the standard formulae

$$
I^{\mathrm{CP}}(\theta^{\mathrm{CP}}) = D^{\top} I^{\mathrm{DP}}(\theta^{\mathrm{DP}})\,D\,, \qquad \mathcal{J}^{\mathrm{CP}}(\hat{\theta}^{\mathrm{CP}}) = \hat{D}^{\top} \mathcal{J}(\hat{\theta}^{\mathrm{DP}})\,\hat{D}\,,
\tag{3.24}
$$

respectively, where D denotes the Jacobian matrix

$$D = (D_{rs}) = \left(\frac{\partial \theta_r^{\mathrm{DP}}}{\partial \theta_s^{\mathrm{CP}}} \right) = \begin{pmatrix} 1 & 0 & -\dfrac{\mu_z}{\sigma_z} & \dfrac{\partial \xi}{\partial \gamma_1} \\ 0 & I_{p-1} & 0 & 0 \\ 0 & 0 & \dfrac{1}{\sigma_z} & \dfrac{\partial \omega}{\partial \gamma_1} \\ 0 & 0 & 0 & \dfrac{d\alpha}{d\gamma_1} \end{pmatrix} \tag{3.25}$$

and $\hat{D}$ denotes D evaluated at the MLE point. The terms in the last column of D are

$$\frac{\partial \xi}{\partial \gamma_1} = -\frac{\sigma \mu_z}{3\sigma_z \gamma_1},$$

$$\frac{\partial \omega}{\partial \gamma_1} = -\frac{\sigma}{\sigma_z^2} \frac{d\sigma_z}{d\alpha} \frac{d\alpha}{d\gamma_1}, \qquad \frac{d\sigma_z}{d\alpha} = -\frac{\mu_z}{\sigma_z} \frac{b}{(1 + \alpha^2)^{3/2}},$$

$$\frac{d\alpha}{d\gamma_1} = \frac{2}{3(4 - \pi)} \left(\frac{1}{T R^2} + \frac{1 - 2/\pi}{T^3} \right), \qquad T = \left[\frac{2}{\pi} - \left(1 - \frac{2}{\pi} \right) R^2 \right]^{1/2}$$

and R is defined in (3.20).

Numerical computation of $\mathcal{I}(\theta^{\mathrm{CP}})$ indicates that this matrix approaches $\mathrm{diag}(1/\sigma^2, 2/\sigma^2, 1/6)$ when the component γ_1 of CP approaches 0. This fact is in agreement with our expectation, since the first two terms coincide with the corresponding terms of a regular normal distribution, and the third term is the inverse asymptotic variance of the sample coefficient of skewness when the data are normally distributed. However, formal computation of the limit of $\mathcal{I}^{\mathrm{CP}}(\theta^{\mathrm{CP}})$ as $\gamma_1 \to 0$ is not amenable to direct computation from the first expression in (3.24), and another route will be described in the next section.

3.1.5 More on the distribution of the MLE

We want to take a closer look at the distribution of the MLE. For simplicity, we confine ourselves to the case of a simple random sample, without covariates, since this is sufficient to illustrate the key concepts.

We have seen that the information matrix (3.18) at $\alpha = 0$ is singular, which violates one of the standard conditions for asymptotic normality of the MLE. This sort of situation falls under the umbrella of the non-standard asymptotic theory developed by Rotnitzky *et al.* (2000) for cases where the information matrix is singular, starting from a motivating problem which has in fact a strong connection with the present setting. The resulting formulation is quite technical and even its use requires some care. An outline of the key aspects is provided by Cox (2006, Section 7.3).

Making use of this theory, various results on the asymptotic distribution of the MLE can be proved. One such finding is that $\hat{\alpha} = O_p(n^{-1/6})$ when $\alpha = 0$, confirming what we have obtained by a direct argument at the beginning of § 3.1.4. Another relevant outcome is to establish that indeed the CP behaves regularly at the point $\gamma_1 = 0$, in two ways: (a) the profile log-likelihood of γ_1 has no stationary point at $\gamma_1 = 0$, and (b) if $\theta^{\rm CP} = (\mu, \sigma, 0)^\top$, then

$$\sqrt{n}\left(\hat{\theta}^{\rm CP} - \theta^{\rm CP}\right) \xrightarrow{\ \mathrm{d}\ } \mathrm{N}_3\left(0, \mathrm{diag}(\sigma^2, \tfrac{1}{2}\sigma^2, 6)\right) \qquad (3.26)$$

which confirms formally the earlier numerical outcome on the information matrix.

So far we have put great emphasis on the subset of the parameter space having $\alpha = 0$, or equivalently $\gamma_1 = 0$. While this case is important, the complement set must be taken into consideration too. If $\alpha \neq 0$, the phenomenon of linear dependence among the components of the score vector (3.2) does not occur. Hence the information matrix is non-singular, and standard results of asymptotic theory apply.

One can however expect that, since $\hat{\alpha} = O_p(n^{-1/6})$ when $\alpha = 0$, this slow convergence of $\hat{\alpha}$ propagates to some extent also to the points nearby, in the sense of slow convergence, as n increases, to the asymptotically normal distribution, and some non-normal behaviour of the MLE at least for small and perhaps even for moderate sample size.

To get a concrete perception of the behaviour of the MLEs, consider the following simulation experiment. A set of 5000 samples of size $n = 200$ each has been generated from an SN(0, 1, 1) variate and, for each sample, the MLE has been obtained. The set of such estimates is represented graphically in Figure 3.4 in the form of a histogram for $\hat{\xi}$ and $\hat{\alpha}$ in the top panels, and scatter plots for the pairs $(\hat{\xi}, \hat{\alpha})$ and $(\hat{\xi}, \hat{\omega})$ in the bottom panels.

The distribution of these estimates is distinctly non-normal, both in the $\hat{\alpha}$ and in the $\hat{\xi}$ component. This outcome is qualitatively in line with the theory of Rotnitzky *et al.* (2000), which indicates that in a neighbourhood of $\alpha = 0$ the estimate $\hat{\alpha}$ can take the wrong sign with probability which can be up to $1/2$, leading to a bimodal distribution, This behaviour is clearly visible in the second histogram of Figure 3.4, which has a mode near $\alpha = 1$ and a secondary mode near $\alpha = -1$. For the reason indicated earlier, the

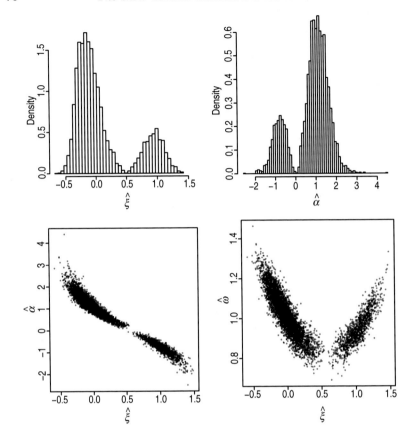

Figure 3.4 Distribution of MLE for samples of size $n = 200$ from SN$(0, 1, 1)$ estimated by simulation of 5000 samples. The top panels represent the histogram of $\hat{\xi}$ (left) and $\hat{\alpha}$ (right); the bottom panels display the scatter plots of $(\hat{\xi}, \hat{\alpha})$ and $(\hat{\xi}, \hat{\omega})$.

bimodal effect is transferred to $\hat{\xi}$ as well. The width of the neighbourhood where the estimate $\hat{\alpha}$ can take on the wrong sign with non-negligible probability decreases to 0 as n diverges, but it is striking to see how long non-normality of the MLE persists, with an appreciable effect even for a parameter value $\alpha = 1$ which is not minute, and for the sample size $n = 200$, which is usually more than adequate to achieve a good agreement with the asymptotic distribution.

For the same set of simulated data, the MLE of CP can be obtained by applying the transformation (3.21) to the 5000 estimates of θ^{DP} computed earlier. The outcome is presented in Figure 3.5 in the form of histograms for $\hat{\mu}$ and $\hat{\gamma}_1$ in the top panels, and scatter plots for the pairs $(\hat{\mu}, \hat{\gamma}_1)$ and $(\hat{\mu}, \hat{\sigma})$

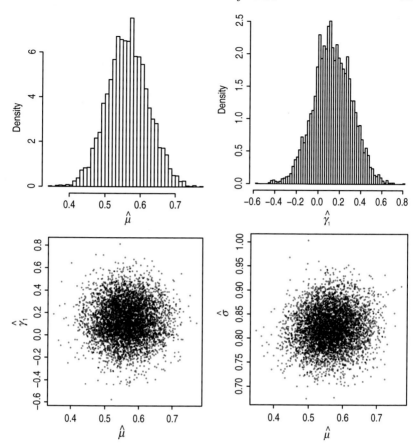

Figure 3.5 Distribution of MLE for samples of size $n = 200$ from SN(0, 1, 1) estimated by simulation of 5000 samples. The top panels represent the histogram of $\hat{\mu}$ (left) and $\hat{\gamma}_1$ (right); the bottom panels display the scatter plots of $(\hat{\mu}, \hat{\gamma}_1)$ and $(\hat{\mu}, \hat{\sigma})$.

in the bottom panels. The CP vector corresponding to $\theta^{\mathrm{DP}} = (0, 1, 1)^{\top}$ is $\theta^{\mathrm{CP}} = (0.564, 0.826, 0.137)^{\top}$.

It is apparent that the distribution of the CP estimates is far preferable as regards closeness to the asymptotic normal distribution. This provides additional evidence that the CP parameterization is far more suitable for the construction of confidence intervals and other methods of inference. In addition to this mathematical aspect, there is a clear advantage in terms of interpretability of CP with respect to DP, as already remarked.

Bibliographic notes

The centred parameterization has been introduced by Azzalini (1985) for simple samples and extended to the regression case by Azzalini and

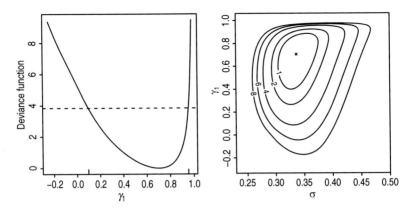

Figure 3.6 Wines data: phenols content of Barolo. Deviance function for the parameter γ_1 of the SN distribution in the left panel and for (σ, γ_1) in the right panel; the mark × indicates $(\hat{\sigma}, \hat{\gamma}_1)$.

Capitanio (1999). The asymptotic distribution (3.26) has been stated without proof by Azzalini (1985) on the basis of numerical evidence and proved formally by Chiogna (2005) using the theory of Rotnitzky *et al.* (2000).

3.1.6 A numerical illustration (continued)

Consider the data introduced in § 3.1.2, specifically the distributions of phenols in Barolo wine, adopting the CP parameterization. Transformation of $\hat{\theta}^{\text{DP}}$ obtained earlier using (3.21) leads to $\hat{\theta}^{\text{CP}} = (2.840, 0.337, 0.703)^{\top}$. These values, especially the first two components, are close to the sample analogues of (μ, σ, γ_1), which are $(2.840, 0.336, 0.795)$; here, the sample standard deviation is uncorrected, as this represents the MLE for normal data.

Figure 3.6 is analogous to Figure 3.3, for CP instead of DP; the first panel refers to γ_1, the second panel to (σ, γ_1). The shape of these curves is free from the kinks observed in Figure 3.3.

The horizontal dashed line in the first panel of Figure 3.6 is at $D = 3.84$, the 0.95-level quantile of the χ_1^2 distribution, and its intersections with the deviance function identify the 95% confidence interval $(0.096, 0.954)$. If this confidence interval is mapped from the γ_1-scale to the α-scale, we obtain the confidence interval $(0.856, 9.75)$ for α. This other interval is the same as obtained if we intersect the deviance function $D(\alpha)$ of Figure 3.3

with the horizontal line at ordinate $D = 3.84$. Notice, however, that the validity of the confidence interval on the α-scale does not follow by application of standard MLE asymptotic theory to $\ell(\theta^{\mathrm{DP}})$, but an asymptotic theory argument applied to θ^{CP} followed by the mapping of the interval from the γ_1-scale to the α-scale.

Similar remarks apply to the (σ, γ_1) profile deviance shown in the right-hand panel of Figure 3.6. Here the region delimited by a given contour level curve can be assigned an approximate confidence level as specified by the χ_2^2 distribution. For instance, the region delimited by the curve labelled 6, which is very close to the 95th percentile of χ_2^2, represents a confidence region of approximate level 0.95. This region, when transferred to the (ω, α) space, via (3.20) and (3.22), corresponds to the region with contour level 6 in the right-hand panel of Figure 3.3, which has then approximate confidence level 0.95 too. Similarly to the earlier case, this statement could not be derived from direct application of standard MLE asymptotic theory to $\ell(\theta^{\mathrm{DP}})$.

This exact correspondence of confidence intervals associated with different parameterizations may not hold if the intervals are produced by a different procedure. Specifically, consider the popular method to produce a confidence interval of level $1 - p$ for a generic parameter θ via the expression $\hat{\theta} \pm z_{p/2} \, \mathrm{std.err.}(\hat{\theta})$ where $z_{p/2}$ is the $(p/2)$-level quantile of the $\mathrm{N}(0, 1)$ distribution. It is well known that in general this procedure lacks equivariance under reparameterization, and the effect can be markedly visible in this context.

An advantage of the CP is that, although not orthogonal, the components of $\hat{\theta}^{\mathrm{CP}}$ appear numerically to be less correlated than those of $\hat{\theta}^{\mathrm{DP}}$. For instance, for the estimates of the distribution in Figure 3.1, inversion of the expected information matrices to get an estimate of the variance matrix of the MLEs, followed by their conversion to correlation matrices, produces

$$\mathrm{cor}\{\hat{\theta}^{\mathrm{DP}}\} \approx \begin{pmatrix} 1 & -0.72 & -0.79 \\ & 1 & 0.71 \\ & & 1 \end{pmatrix}, \quad \mathrm{cor}\{\hat{\theta}^{\mathrm{CP}}\} \approx \begin{pmatrix} 1 & 0.47 & 0.03 \\ & 1 & 0.38 \\ & & 1 \end{pmatrix}.$$

Reduced correlation is a convenient aspect for the interpretation of parameters, which has been observed in several other numerical cases.

Another aspect of the CP parameterization to be illustrated refers to the regression context. For simplicity, assume that only one covariate x is related to a response variable y via

$$y_i = \beta_0 + \beta_1 \, x_i + \varepsilon_i, \qquad i = 1, \ldots, n \tag{3.27}$$

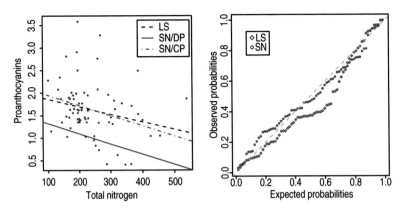

Figure 3.7 Wines data: proanthocyanins versus total nitrogen in Grignolino wine. The left panel displays the data scatter with superimposed the regression line fitted by least squares and by ML estimation under SN assumption, using DP and CP. The right panel displays the PP-plot diagnostics for the two fitted models.

where $\varepsilon_i \sim SN(0, \omega^2, \alpha)$, with independence among different (x_i, y_i)'s. Equivalently, we say that y_i is sampled from $SN(\xi_i, \omega^2, \alpha)$ where $\xi_i = \beta_0 + \beta_1 x_i$.

To illustrate the idea, we make use again of the wines data described earlier, but this time we use a different cultivar, Grignolino, so now there are $n = 71$ specimens; the variables under consideration are total nitrogen (x) and proanthocyanins (y). The data scatter is shown in the left panel of Figure 3.7, with the least-squares line (dashed black) and two red lines for the SN fit superimposed. The solid red line corresponds to ML estimates of β_0 and β_1 as indicated in (3.27), but clearly this line falls far too low in the cloud of points, because we are not taking into account that $\mathbb{E}\{\varepsilon\} \neq 0$. Adjusting β_0 to $\beta_0 + \mathbb{E}\{\varepsilon\}$ amounts to considering the CP intercept (3.23). This correction produces the red dashed line, parallel to the earlier one, now interpolating the points satisfactorily.

For completeness we report below the estimates and standard errors for the two parameter sets, DP and CP. Since β_1 is in common, it is reported only once.

	β_0^{DP}	β_1	ω	α	β_0^{CP}	σ	γ_1
estimate	1.547	−0.00228	0.853	2.49	2.179	0.573	0.57
std. err.	0.203	0.00089	0.110	0.96	0.223	0.052	0.20

To compare the adequacy of the fitted model under the normal and the SN assumption, we make use of the diagnostics introduced in § 3.1.2. In this case we adopt a variant form with respect to Figure 3.2 obtained by transforming the plotted quantities to the probability scale, using the χ_1^2 distribution function. Hence both axes in the right panel of Figure 3.7 range between 0 and 1, and this diagnostic is called a PP-plot.

To ease comparison of the two fits, in this case we have plotted both sets of points on the same diagram. It is visible that the points of the SN residuals are more closely aligned along the identity line than the other set, indicating a better fit to the data. Using a PP-plot in this example and a QQ-plot in Figure 3.2 has no special meaning; both types of plots could have been used in both examples. Note that the residuals $\hat{z}_i$ used in these diagnostic plots are computed from (3.12), irrespective of the parameterization adopted for inference, DP or CP.

3.1.7 Computational aspects

It has already been remarked that the actual computation of the MLE requires numerical techniques, because of the non-linear function ζ_1 appearing in the likelihood equations (3.3). Since the popular statistical computing environment R (R Development Core Team, 2011) provides more facilities for numerical optimization than for solution of non-linear equations, we consider direct maximization of the log-likelihood function. In fact this is the route taken by the R package sn, which is the tool used for the numerical work of this book. However, a good deal of the considerations which follow are useful also for alternative computational routes, such as numerical solution of the likelihood equations.

To choose a starting point for the numerical search, a quite natural option is offered by the method of moments. In the case of a simple sample, denote by $\bar{y}$, s and $\bar{\gamma}_1$ the sample version of the mean, the standard deviation and the coefficient of skewness, respectively. These are taken as estimates of the corresponding CP components (3.21). To convert these CP estimates to DP value, use of (3.20) with γ_1 replaced by $\bar{\gamma}_1$ provides an estimate of α, say $\bar{\alpha}$. Plugging $\bar{y}$, s and $\bar{\alpha}$ in (3.22), we obtain estimates of the other two components, $\bar{\omega}$ and $\bar{\xi}$ say.

This scheme assumes that $\bar{\gamma}_1$ belongs to the admissible interval given by the first expression of (2.30), but of course in practical cases this condition may not hold true. This is why the method of moments is not a viable general methodology in this context. However, for the purpose of selecting initial values of a maximum likelihood search, it is legitimate to replace an observed $\bar{\gamma}_1$ outside the admissible interval by a value just inside the interval, and then proceed as indicated above.

In a regression problem, the μ component of CP is replaced by a vector β whose initial estimate can be provided by least squares. Conversion to the DP scale requires adjustment only of the first component of β, as indicated by (3.23). The adjustment merely requires subtraction of $b\,\bar{\omega}\,\delta(\bar{\alpha})$, which is the same term entering $\bar{\xi}$ above.

Starting from these initial values, a search of the DP parameter space is performed to maximize the log-likelihood. The process can be speeded up, often considerably, if the first derivatives (3.2) and (3.9) are supplied to the optimization algorithm. Further improvement can be achieved by making use also of the second derivatives (3.15). In the alternative problem of solving the likelihood equations, (3.15) provides the derivatives of the score functions to be equated to 0.

Once the DP estimate $\hat{\theta}^{\mathrm{DP}}$ has been obtained, this is simply mapped to the CP space using (3.21) to get $\hat{\theta}^{\mathrm{CP}}$, recalling the equivariance property of MLE. Standard errors are computed via either form in (3.24), but the second one is usually regarded as preferable in likelihood-based inference.

A variant of this scheme is to perform the optimization search directly over the CP space, because of the more regular behaviour of the log-likelihood function. This choice involves computing, for each searched point of the CP space, the scores in the corresponding DP space and their transformation to CP scores via (3.25). A similar transformation is required for the second derivatives, essentially using (3.15) and the first of (3.24). This is the strategy adopted by the R package sn for this MLE problem.

A general problem with maximum likelihood estimation is to determine whether the parameter value selected by the optimization routine corresponds to a global or to a local maximum, apart from a limited set of estimation problems where uniqueness of the maximum can be established on a theoretical basis. In the present context, existence of multiple local maxima is possible, as demonstrated by the sample with $n = 20$ reported by Pewsey (2000a), where two local maxima exist in the interior of the parameter space. However, extensive numerical exploration has indicated that cases of this sort are unusual even for small samples of size $n = 20$ and their frequency of occurrence vanishes rapidly as n increases.

Another approach to maximization of the likelihood is via the EM algorithm or some of its variants. A formulation of this type is quite natural in the present context, if one recalls the stochastic representations (2.13) and (2.14), since they both involve latent components which, if observed, would lead to a manageable estimation problem. For more in this direction see Complement 3.1 and Problem 3.4. However, the fact that conceptually an EM algorithm fits well in this context does not necessarily entail a superior numerical performance with respect to direct maximization of the log-likelihood.

3.1.8 Boundary estimates

Another peculiar behaviour of the SN log-likelihood function is that, in some cases, it does not have a maximum in the interior of the parameter space. Liseo (1990) has examined this phenomenon in the one-parameter case when a random sample $z = (z_1, \ldots, z_n)$ is drawn from $Y \sim SN(0, 1, \alpha)$. The likelihood function

$$L(\alpha) = \text{constant} \times \ \Phi(\alpha z_1) \times \ \cdots \times \ \Phi(\alpha z_n)$$

can be a strictly monotonic function, if it happens that all sample values have the same sign. This implies that the MLE is $\hat{\alpha} = \pm\infty$, where the sign is the same as for the z_i's; equivalently, the MLE of γ_1 is $\hat{\gamma}_1 = \pm\gamma_1^{\max}$.

With the aid of Proposition 2.7(d), the probability of incurring a sample of the above pattern can easily be computed to be

$$p_{n,\alpha} = \left(\frac{1}{2} - \frac{\arctan \alpha}{\pi} \right)^n + \left(\frac{1}{2} + \frac{\arctan \alpha}{\pi} \right)^n ,$$

which goes to 0 as $n \to \infty$, provided $|\alpha| < \infty$, but for finite sample size it can be appreciable. For instance, if $\alpha = 5$ and $n = 25$, $p_{n,\alpha}$ is about 0.20; from here it drops rapidly when n increases: keeping $\alpha = 5$, $p_{n,\alpha} < 0.04$ if $n = 50$, and $p_{n,\alpha} < 10^{-7}$ if $n = 250$.

If a location and a scale parameter are inserted in the model, so that we are back to the log-likelihood with elements of type (3.1), MLEs on the boundary of the parameter space still occur, but it is not currently known which data patterns lead to this outcome. An illustrative example is displayed in Figure 3.8, which refers to an artificial sample of size $n = 50$ sampled from $SN(0, 1, 5)$. The individual data points are indicated by the ticks at the bottom, the solid curve denotes the true density, the dot-dashed line corresponds to the MLE, and the dashed line is a non-parametric estimate of the density. The MLE has $\hat{\alpha} = \infty$ and $\hat{\xi}$ is just below the smallest observation.

This numerical outcome raises the question of how to handle this type of situation. Two essentially opposite attitudes are as follows.

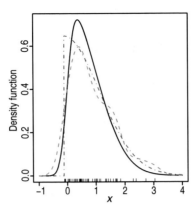

Figure 3.8 A sample of size $n = 50$ from SN(0, 1, 5), whose
density function is denoted by a solid line, leading to MLE on the
boundary of the parameter space (dot-dashed line) with a
non-parametric density estimate (dashed line) superimposed.

◇ One way of looking at it is to regard this numerical result on the footing of any other point of the parameter space. In this case, the MLE
has just happened to land on the boundary of the parameter space, a
somewhat unusual distribution in the present setting, with support on a
subset of the real line, but still a legitimate member of the SN parametric
family. To draw an analogy with the more familiar case of independent
Bernoulli trials, this outcome is similar to the case when all observations are 0; the estimated probability of success is then $\hat{p} = 0$, even if
this corresponds to a degenerate binomial distribution. Notice, however,
that similarly to Bernoulli trials with $\hat{p} = 0$, this approach still requires
a special treatment for other inferential aspects, especially for interval
estimation, since standard MLE asymptotic theory does not deal with
the case of boundary points of the parameter space.

◇ An alternative view of the problem is to reject the boundary estimate as
a valid one, since this can happen with samples whose patterns are not in
agreement with the distribution associated with $\hat{\alpha} = \pm\infty$. For instance,
the data of Figure 3.8 do not have a pattern of decreasing density on the
right of the minimum observation, as $\hat{\alpha} = \infty$ is saying. This is visible by
direct inspection of the data, and it is more conveniently indicated by the
non-parametric estimate. In addition, the sample coefficient of skewness
is $\bar{\gamma}_1 = 0.9022$, and corresponds to $\bar{\alpha} = 6.38$, using (3.20); this value
is inside the parameter space, providing another element of evidence for

a non-degenerate sample. The situation is therefore very different from the case of Bernoulli trials, where $\hat{p} = 0$ can be obtained only by a set of data all pointing in that direction.

The second approach, which effectively amounts to dropping $\alpha = \pm\infty$ from the admissible set, is the only one which has actually been developed so far. Among the various proposals that have been examined, those more closely linked to the classical inferential paradigm will be presented in the rest of this section.

Sartori (2006) has proposed a method to avoid boundary estimates based on a modification of the likelihood equations which Firth (1993) had put forward as a general bias-reduction technique. In fact, the occurrence of $|\hat{\alpha}| = \infty$ with non-null probability produces maximal bias. Phrased in the one-parameter case, θ say, Firth's method replaces the usual score function $S(\theta)$ by a modified form $S^*(\theta)$ with corresponding modified likelihood equation

$$S^*(\theta) = S(\theta) - I(\theta)\,b(\theta) = 0, \tag{3.28}$$

where $b(\theta)$ is the leading term of bias of the MLE, typically $O(n^{-1})$, and $I(\theta)$ is the expected Fisher information. The extra term $-I(\theta)\,b(\theta)$ is chosen in such a way that the estimate obtained by solving (3.28) has a bias of order of magnitude $O(n^{-2})$.

If this scheme is applied to the case of a random sample $z = (z_1, \ldots, z_n)^\top$ from SN$(0, 1, \alpha)$, (3.28) takes the form

$$\sum_{i=1}^n \zeta_1(\alpha z_i)\,z_i - \frac{\alpha}{2}\frac{a_4(\alpha)}{a_2(\alpha)} = 0, \tag{3.29}$$

where $a_k(\alpha)$ is given by (3.16); the first term of (3.29) is as in the third equation (3.3). Sartori has proved that (3.29) always admits a finite solution; he did not prove that there is a unique solution, although this was true in all cases examined numerically.

An interesting aspect of the adjustment term of the score function in (3.28) is that, in full exponential families, it coincides with the term produced, in a Bayesian context, by the adoption of Jeffreys' prior distribution for θ. Since (3.29) does not arise from an exponential family, this equality does not hold exactly, but Sartori notes that the shape of the two functions is similar.

A difficulty with this approach is that it is not easily implementable in more complex situations, such as the three-parameter case where $\theta = (\xi, \omega, \alpha)^\top$ or the more elaborate formulations examined in subsequent chapters, because of the difficulty of obtaining the analytical expressions of the adjustment term in (3.28). Moreover, even for the simple case of (3.29), there is the practical disadvantage that each function evaluation requires computing two integrals numerically, $a_2(\alpha)$ and $a_4(\alpha)$.

These facts motivate a related but somewhat distinct formulation put forward by Azzalini and Arellano-Valle (2013), who consider the *penalized log-likelihood*

$$\ell_p(\theta) = \ell(\theta) - Q(\theta), \qquad (3.30)$$

where θ denotes the set of parameters in the setting under consideration and the penalty function $Q(\theta)$ satisfies

$$Q(\theta) \ge 0, \qquad Q(\theta)\big|_{\alpha=0} = 0, \qquad \lim_{|\alpha|\to\infty} Q(\theta) = +\infty \qquad (3.31)$$

and $Q(\theta)$ does not depend on n. It would be possible to allow $Q(\theta)$ to depend on the y values, provided it remains 'bounded in probability' as n diverges, that is $Q(\theta) = O_p(1)$. However, this variant is not necessary in the specific construction below.

Under these assumptions, $\ell_p(\theta)$ takes its maximum value at a finite point, $\tilde{\theta}$ say, which we shall denote the maximum penalized likelihood estimate (MPLE). Under the above conditions plus some other standard regularity conditions, it is easy to show that $\tilde{\theta}$ and $\hat{\theta}$ differ by a vanishing amount as n diverges. This implies that asymptotic distributional properties $\tilde{\theta}$ are the same as $\hat{\theta}$ in the first-order approximation. Standard errors can be obtained from the corresponding penalized information matrix via

$$\mathrm{var}\{\tilde{\theta}\} \approx -\ell_p''(\tilde{\theta})^{-1}. \qquad (3.32)$$

Conditions (3.31) on $Q(\theta)$ leave ample room for choice. To narrow down the choice, notice that the bias correction factor in (3.28) plays the role of the derivative of (3.30). Therefore, in the one-parameter case where θ is α, we can take $Q'(\alpha)$ equal to the correction factor in (3.29). Moreover, we can exploit a convenient fact illustrated in the left plot of Figure 3.9, where the circles represent the ratios $a_2(\alpha)/a_4(\alpha)$, evaluated by numerical integration, plotted versus α^2 for a range of values of α. The beautiful alignment of these points makes it almost compelling to interpolate them linearly. The coefficients of the line can be chosen by matching the exact values at $\alpha^2 = 0$ and $\alpha^2 \to \infty$. To this end, rewrite $a_k(\alpha)$ defined by (3.16) as

$$a_k(\alpha) = \sqrt{\frac{2}{\pi}} \frac{1}{(1+\alpha^2)^{(k+1)/2}} \, \mathbb{E}\{X^k \, \zeta_1(\delta X)\}, \qquad (3.33)$$

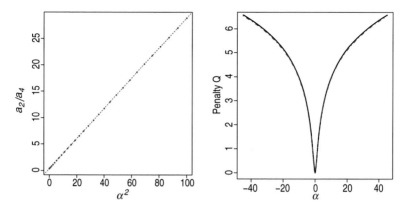

Figure 3.9 The points in the left panel represent the exact values of $a_2(\alpha)/a_4(\alpha)$ plotted versus α^2, for a selection of α values, with superimposed interpolating line having intercept e_1 and slope e_2. The right panel compares the exact integral (solid curve) of the penalty in (3.29) with its approximation (3.35) (dot-dashed curve).

where $X \sim N(0, 1)$ and $\delta = \delta(\alpha) = \alpha/\sqrt{1 + \alpha^2}$. Hence write

$$\frac{a_2(\alpha)}{a_4(\alpha)} = (1 + \alpha^2)\frac{\mathbb{E}\{X^2\, \zeta_1(\delta X)\}}{\mathbb{E}\{X^4\, \zeta_1(\delta X)\}} \approx e_1 + e_2\, \alpha^2, \qquad (3.34)$$

leading to

$$e_1 = \frac{a_2(0)}{a_4(0)} = \frac{\mathbb{E}\{X^2\}}{\mathbb{E}\{X^4\}} = \frac{1}{3},$$

$$e_2 = \lim_{\alpha^2 \to \infty}\left(\frac{1 + \alpha^2}{\alpha^2}\frac{\mathbb{E}\{X^2\, \zeta_1(\delta X)\}}{\mathbb{E}\{X^4\, \zeta_1(\delta X)\}} - \frac{e_1}{\alpha^2}\right) = \frac{\mathbb{E}\{X^2\, \zeta_1(X)\}}{\mathbb{E}\{X^4\, \zeta_1(X)\}} \approx 0.2854166,$$

where the final coefficient has been obtained by numerical integration. The dashed line in the left panel of Figure 3.9 has intercept e_1 and slope e_2.

On replacing $a_2(\alpha)/a_4(\alpha)$ by $e_1 + e_2\alpha^2$ in $Q'(\alpha)$ and integrating, we get

$$Q = c_1 \log(1 + c_2\, \alpha^2) \qquad (3.35)$$

where $c_1 = 1/(4\,e_2) \approx 0.87591$ and $c_2 = e_2/e_1 \approx 0.85625$. The right plot of Figure 3.9 compares (3.35) with the numerical integral of the correction factor in (3.29), shifted to have minimum value 0, confirming the accuracy of the approximation.

In the multi-parameter case where α is only a component of θ, it is sensible to penalize the log-likelihood function only on the basis of the α component. Hence we keep using (3.35) as the Q function of (3.30). Simulation work of Azzalini and Arellano-Valle (2013) indicates that this option produces a reasonable outcome in the three-parameter case also. For the data of Figure 3.8, the MPLE of $\theta = (\xi, \omega, \alpha)^{\top}$ is $(0.034, 1.165, 6.256)^{\top}$, not far from the true value $(0, 1, 5)^{\top}$.

An alternative approach to the problem of boundary estimates has been proposed by Greco (2011). The method is based on the idea of minimizing the Hellinger distance between the density corresponding to a given choice of the parameters and a non-parametric density estimate. The author proves that the method delivers finite estimates of the slant in the one-parameter case and that the estimates are asymptotically fully efficient.

3.2 Bayesian approach

Consider first the one-parameter case where observations are sampled from $SN(0, 1, \alpha)$. Jeffreys' prior distribution $\pi_J(\alpha)$ is proportional to the square root of the element in the bottom-right corner of (3.17), that is

$$\pi_J(\alpha) \propto a_2(\alpha)^{1/2} = \tilde{\pi}_J(\alpha), \qquad \alpha \in \mathbb{R}, \tag{3.36}$$

say, where a_2 is given by (3.16). After rewriting $a_2(\alpha)$ as

$$\begin{aligned} a_2(\alpha) &= \int_{-\infty}^{\infty} 2 z^2 \varphi(z) \frac{\varphi^2(\alpha z)}{\Phi(\alpha z)} dz \\ &= \int_0^{\infty} 2 z^2 \varphi(z) \frac{\varphi^2(\alpha z)}{\Phi(\alpha z)} dz + \int_0^{\infty} 2 z^2 \varphi(z) \frac{\varphi^2(\alpha z)}{\Phi(-\alpha z)} dz \\ &= \int_0^{\infty} 2 z^2 \varphi(z) \frac{\varphi^2(\alpha z)}{\Phi(\alpha z)\Phi(-\alpha z)} dz \\ &= a_2(-\alpha), \end{aligned}$$

Liseo and Loperfido (2006) show that a_2 decreases monotonically with $|\alpha|$, with tails of order $O(|\alpha|^{-3})$ for large $|\alpha|$. This implies that $\tilde{\pi}_J(\alpha)$ is integrable, providing one of the rare examples where the Jeffreys' prior of a parameter with unbounded support is a proper probability distribution.

The logarithm of the posterior distribution, given an observed sample $z = (z_1, \ldots, z_n)^{\top}$, is of the form

$$\log \pi(\alpha|z) = \text{constant} + \log L(\alpha; z) + \log \tilde{\pi}_J(\alpha),$$

where the log-likelihood $\log L(\alpha; z) = \sum \zeta_0(\alpha z_i)$ is bounded from above and $\log \tilde{\pi}_J(\alpha) \to -\infty$ as $|\alpha| \to \infty$. This ensures that the mode of the posterior distribution is always finite.

In the three-parameter case, another result of Liseo and Loperfido (2006) provides, up to a normalization constant, an explicit expression of the integrated likelihood for α. Unfortunately, this involves the n-dimensional Student's t distribution function, which becomes rapidly untractable as n increases beyond a few units. Therefore, its use is more for additional theoretical developments than for actual practical work.

To overcome these problems, Bayes and Branco (2007) make use of the approximation

$$\frac{1}{\pi} \frac{\varphi(x)}{\sqrt{\Phi(x)\Phi(-x)}} \approx b^2 \varphi(b^2 x), \tag{3.37}$$

where $b = \sqrt{2/\pi}$. This arises by first noticing that graphically the term on the left has the same behaviour as a $N(0, \sigma^2)$ density; then choose the scale factor $\sigma = b^{-2}$ by matching the two expressions at $x = 0$. Using (3.37) in the above expression for $a_2(\alpha)$, one arrives at the approximation

$$a_2(\alpha) \approx \frac{b^2}{(1 + 2 b^4 \alpha^2)^{3/2}}$$

which turns out to be numerically quite accurate. From (3.37), approximations for other terms of type (3.16) can also be produced. The square root of the above expression is proportional to the density of a Student's t variate with $\frac{1}{2}$ degrees of freedom, multiplied by the scale factor b^{-2}. On insertion of the appropriate normalizing constant, we obtain that the Jeffreys' prior can be closely approximated as

$$\pi_J(\alpha) \approx \frac{\Gamma(3/4)}{\Gamma(1/4)} \frac{b^3}{(1 + 2 b^4 \alpha^2)^{3/4}}. \tag{3.38}$$

Note that $-\log \pi_J(\alpha)$ is of type (3.35), up to an irrelevant additive constant, with coefficients numerically close to c_1 and c_2. This can also be confirmed by a plot of $-\log \pi_J(\alpha)$ versus α, which would look much the same as the right-side panel of Figure 3.9.

Another way to introduce a vague prior distribution for the slant parameter is to adopt a uniform distribution for δ over the interval $(-1, 1)$. The transformation from δ to $\alpha = \delta/\sqrt{1 - \delta^2}$ produces the density

$$\frac{1}{2 (1 + \alpha^2)^{3/2}}, \qquad \alpha \in \mathbb{R},$$

which is qualitatively similar to $\pi_j(\alpha)$, since it is the density of a Student's t_2 variable multiplied by a factor $1/\sqrt{2}$.

Therefore, in both cases, α can be represented as a normal variate with random scale factor S such that S^{-2} is a suitably scaled Gamma variate, with shape index either $1/4$ or 1. Bayes and Branco (2007) combine this fact with the stochastic representation (2.14) and a standard assumption of improper prior distribution proportional to ω^{-1} for the location and scale parameters, ξ and ω. After suitable reparameterization, this leads to a Gibbs sampling mechanism which allows us to estimate by simulation the posterior distribution of the parameters of interest.

Simulation work of Bayes and Branco (2007) indicates a good performance of the posterior mode starting from the Jeffreys' prior as an estimate of α, and indicates that this mode is very close to Sartori's estimate. For interval estimation, high posterior density regions based on the uniform prior for δ are somewhat preferable.

An alternative form of 'objective' Bayesian analysis has been developed by Cabras *et al.* (2012) making use of the idea of a 'matching prior', that is, 'a prior for which Bayesian and frequentist inference agree to some order of approximation'. In our context, the matching prior for α, allowing for the presence of $\psi = (\xi, \omega)$, is

$$\pi_m(\alpha) \propto \left. \left(\mathcal{I}_{\alpha\alpha}(\theta^{\mathrm{DP}}) - \mathcal{I}_{\alpha\psi}(\theta^{\mathrm{DP}}) \, \mathcal{I}_{\psi\psi}(\theta^{\mathrm{DP}})^{-1} \, \mathcal{I}_{\psi\alpha}(\theta^{\mathrm{DP}}) \right)^{1/2} \right|_{\theta^{\mathrm{DP}} = (\hat{\psi}(\alpha), \alpha)}, \qquad (3.39)$$

where the terms involved are the blocks of the information matrix (3.17), having dropped the superscript from $\mathcal{I}^{\mathrm{DP}}$ for simplicity. Since ξ does not appear in the information matrix and the terms ω cancel out, (3.39) is a function of α only, not depending on the data.

An explicit expression for $\pi_m(\alpha)$ is then available, up to the terms a_0, a_1, a_2 involved in (3.17). This expression is not particularly appealing, but its graphical appearance is visible in Figure 3.10, superimposed on π_j. It can be established that $\pi_m(\alpha)$ is symmetric about 0 and decreases at a rate $|\alpha|^{-3/2}$ as $\alpha \to \pm\infty$, implying that the function is integrable. Therefore, similarly to π_j, π_m is also a proper symmetric density function. A distinctive difference from (3.38) is that $\pi_m(\alpha)$ does not have a mode at 0; on the contrary, there is an antimode: $\pi_m(0) = 0$.

A simulation experiment carried out by Cabras *et al.* (2012) confirms that the use of π_m leads to credibility intervals with a closer agreement between nominal and actual coverage probabilities compared to π_j. This is as expected, given the principle which drives the construction of π_m.

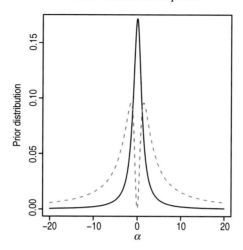

Figure 3.10 Two forms of 'objective' prior distribution for the parameter α of a skew-normal distribution: the solid line is the Jeffreys' prior, the dashed line is the matching prior.

Another advantage of the matching prior is that it extends in a simple manner to the linear regression case, without additional computational burden.

From the classical viewpoint, the fact that $\pi_m(0) = 0$ could be exploited to rule out estimates at $\alpha = 0$, which is a problematic value for inference. This would amount to setting $Q = -\log \pi_m(\alpha)$ in (3.30), a choice that effectively removes $\alpha = 0$ from the parameter space, besides $\alpha = \pm\infty$. Correspondingly, the second requirement of (3.31) must be changed to $Q(0) = \infty$. Since the new Q is still bounded in probability, the argument leading to (3.32) would still apply, provided the condition $\alpha \neq 0$ holds true.

3.3 Other statistical aspects

3.3.1 Goodness of fit

We examine two conceptually distinct but not unrelated problems. One question is as follows: for a given set of data, $y = (y_1, \ldots, y_n)^\top$, is the classical normal family of distributions adequate to describe their distribution, or do we need to introduce a more flexible one, such as the skew-normal? Here the question of adequacy of the normal distribution is raised without specification of its parameters, mean and variance.

To tackle this problem in § 3.1.2, we have adopted a simple procedure based on the sample coefficient of skewness, $\bar{\gamma}_1$. After suitable standardization with its asymptotic standard deviation $\sqrt{6/n}$ or with the exact standard deviation under normality, a formal test procedure can be formulated, by comparing the standardized value of $\bar{\gamma}_1$ with the interval associated with the percentage points of the $N(0, 1)$ distribution at the selected significance level; equivalently, but operationally preferably, the observed significance level can be computed.

This time-honoured tool, $\bar{\gamma}_1$, has a special appeal in the present context, since Salvan (1986) has shown that the locally most powerful location-scale invariant test, when the null hypothesis is represented by the normal family and the alternative is formed by the skew-normal set of distributions with $\alpha > 0$ or equivalently $\gamma_1 > 0$, is based on the test statistic $\bar{\gamma}_1$. Recall from § 3.1.1 that it is not feasible to test for normality within the SN class on the basis of the score function for the direct parameters evaluated at $\alpha = 0$, and the development of Salvan (1986) involves consideration of the third derivative of a suitably defined marginal invariant likelihood.

The second question of interest is this: given a set of data, does the skew-normal distribution provide an adequate probability model for the generating mechanism of the data? Again the question is raised without specification of the value of (ξ, ω, α). This requirement produces a serious complication since in this case there is no known transformation of the data whose distribution is parameter invariant, such as (3.14) in the normal case.

To tackle this problem, Dalla Valle (2007) has proposed a testing procedure based on the following main steps. First, the MLE of $\hat{\theta}^{\mathrm{DP}} = (\hat{\xi}, \hat{\omega}, \hat{\alpha})^{\top}$ is estimated from the sample y, leading to the corresponding values of the integral transform $u_i = \Phi(z_i; \hat{\alpha})$, where z_i is as in (3.12), for $i = 1, \dots, n$. Next, for these u_i's, the Anderson–Darling test statistic is computed to test for a uniform distribution, which would hold exactly if the z_i's were computed using the true parameter value. Because of the replacement of θ^{DP} by $\hat{\theta}^{\mathrm{DP}}$, the null distribution of the test statistic differs from the nominal one associated with the exact $U(0, 1)$ distribution of the u_i's. For a selected significance level of the test, typically 5%, an approximate percentage point is computed as a function of $\hat{\alpha}$ and n, and the observed value of the Anderson–Darling statistic is compared with this approximate quantile. Three such functions are considered, but one of them, denoted 'minimum value', is recommended as preferable. Simulation results indicate that the actual significance level of this procedure, as α ranges from 1 to 20, is acceptably close to the nominal one. For instance, if $n = 100$, the nominal significance level 5% corresponds to an actual level with ranges between 4.3% and 6.3%.

Other testing procedures for this problem have been put forward by Mateu-Figueras *et al.* (2007), Meintanis (2007), Cabras and Castellanos (2009) and Pérez Rodríguez and Villaseñor Alva (2010).

Alternatively to the above formal test procedures, one can resort to a graphical procedure, based on the approximate χ_1^2 distribution of the square of the residuals (3.12), already described in § 3.1.2 and illustrated by Figure 3.2(a) and Figure 3.7(b), in the QQ-plot and PP-plot form, respectively.

3.3.2 Inference for the ESN family

Exploratory numerical work fitting the ESN family to some data set has been done by Arnold *et al.* (1993, Section 6). However, they report that the addition of the fourth parameter τ to (ξ, ω, α) causes 'severe identifiability problems', and caution against the use of this distribution for data fitting, unless τ is known. Additional numerical work of Capitanio *et al.* (2003, Section 4) and Canale (2011) gives similar indications.

A qualitative explanation of these problems with the ESN family is suggested by Figure 2.4, where most of the curves appear to behave quite similarly to members of the SN family; see Figure 2.1 for a comparison. Only for a few combinations of α and τ do there appear to be some visible differences between the curves of Figure 2.4 and the SN densities, and even then only for a limited range of the abscissa. The indication is that, for a combination of parameters $(\xi, \omega, \alpha, \tau)$, there will be a member of the ESN family with parameters $(\xi', \omega', \alpha', 0)$ whose density function is about the same.

A formal explanation of the behaviour described above has been provided by Canale (2011), who has obtained an explicit expression for the 4-dimensional expected Fisher information matrix for the ESN family. This matrix is singular when $\alpha = 0$, similarly to the SN case, and here τ becomes exactly not identifiable. For $\alpha \neq 0$, the matrix is non-singular, but its determinant is small and in fact extremely small for most of the (α, τ) space. This determinant is represented, for the case of a single observation, in two graphical forms in Figure 3.11; here the scale parameter ω has been set to 1, and the value of ξ does not affect the matrix.

The maximum value achieved by the determinant is about 3.5×10^{-4}, which is very small compared with, say, the corresponding value for (μ, σ) of a normal distribution when $\sigma = 1$, in which case the determinant is constantly 2. Moreover, as soon as we move a little away from the area

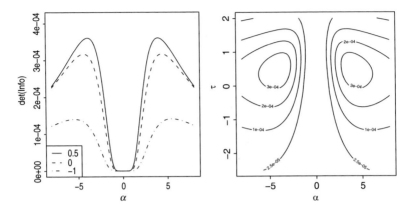

Figure 3.11 Determinant of the expected Fisher information matrix for a single observation from an ESN variable when $\omega = 1$. In the left panel, the determinant is plotted versus α for a few selected values of τ; in the right panel, the determinant is represented as contour level curves of (α, τ).

of maximal information, the determinant decreases rapidly, and it is essentially 0 over a vast area of the parameter space.

If all four parameters $(\xi, \omega, \alpha, \tau)$ have to be estimated, it is advisable to tackle maximization of the log-likelihood as a sequence of three-parameter estimation problems for a range of fixed values of τ, and select the value $\hat{\tau}$ of τ with largest log-likelihood, and the corresponding values of the other parameters. In other words, one constructs the profile log-likelihood function for τ. This scheme offers some improvement in numerical stability, but it cannot overcome the problems intrinsic to the formulation. Hence, the profile log-likelihood can happen to be very flat, and in some cases it is monotonic, so that $\hat{\tau}$ diverges, more easily to $-\infty$.

These points are illustrated by Figure 3.12, which displays the profile deviance function of τ for the data presented in § 3.1.2. Although a finite MLE of τ exists at about 0.6, any other value on its left is essentially equivalent from a likelihood viewpoint. In addition, the curve is grossly irregular near $\tau = 2$. All of this confirms the 'severe identifiability problems' quoted at the beginning of this section.

The above discussion prompts two general remarks, pointing to opposite directions. On the one side, this case illustrates how the introduction of more and more general families of distributions does not automatically translate into an improvement from a statistical viewpoint. Introducing additional parameters may actually complicate the inferential process, when

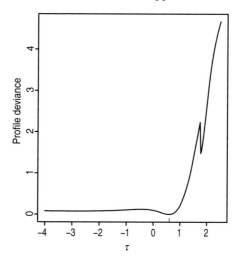

Figure 3.12 Total contents of phenols in Barolo wine: profile deviance function of τ.

they do not lead to a relevant flexibility of the parametric family, so that the net effect of their introduction may actually be negative. This is an instance of the potential source of problems referred to in the cautionary note at the end of § 1.2.1.

On the other side, the difficulties in fitting this distribution to observed data do not imply that the ESN family is useless in practice, but care is required, as well as an adequate sample size. In addition, its relevance for purposes different from data fitting will emerge in Chapter 5.

3.4 Connections with some application areas

3.4.1 Selective sampling and partial observability

In social and economic studies, potential bias in sample selection is a crucial theme, because of the observational nature of many studies. This problem has motivated a vast literature, whose common feature can be described as partial observability of dependence models, due to some form of censoring or truncation of the response variable. A very influential paper in this area is by Heckman (1976); for a concise account providing basic information about several related directions of work, see Maddala (2006). An extensive discussion from a statistical viewpoint has been provided by Copas and Li (1997).

A simple scheme which retains the essential ingredients of more complex formulations involves two regression models of the form

$$
\begin{aligned}
Y &= x^\top \beta + \sigma \varepsilon_1, \\
W &= w^\top \gamma + \varepsilon_2,
\end{aligned}
\tag{3.40}
$$

where x and w are vectors of explanatory variables which we shall regard as fixed, β and γ are vector parameters, σ is a positive scale parameter, and $\varepsilon = (\varepsilon_1, \varepsilon_2)^\top$ is a bivariate normal variable with standardized marginals and $\mathrm{cor}\{\varepsilon_1, \varepsilon_2\} = \delta$. The first regression model in (3.40) is the element of interest, but we do not observe Y and W directly. Only the indicator variable $W^* = I_{(0,\infty)}(W)$ is recorded from W, and Y is observed only for the individuals with $W > 0$.

When $\delta = 0$, observation or censoring of Y occurs completely at random, and the only effect of the condition $W > 0$ is a reduction of the sample size. For general δ, selective sampling takes place, which prevents appropriate use of least squares and other standard methods, and has prompted the development of the above-mentioned literature. Here we shall confine ourselves to highlighting the connections with our treatment.

From representation (2.41), we see that the distribution of Y conditionally on $W > 0$ is of ESN type, specifically

$$
(Y|W > 0) \sim \mathrm{SN}(x^\top \beta, \sigma^2, \alpha(\delta), x^\top \gamma),
$$

and consequently the mean value and the variance of the observable Y's are

$$
\begin{aligned}
\mathbb{E}\{Y|W > 0\} &= x^\top \beta + \sigma \, \mathbb{E}\{\varepsilon_1|\varepsilon_2 + x^\top \gamma > 0\} = x^\top \beta + \sigma \, \delta \, \zeta_1(x^\top \gamma), \\
\mathrm{var}\{Y|W > 0\} &= \sigma^2 \, \mathrm{var}\{\varepsilon_1|\varepsilon_2 + x^\top \gamma > 0\} = \sigma^2 \{1 + \delta^2 \, \zeta_2(x^\top \gamma)\},
\end{aligned}
$$

respectively. These expressions match analogous ones of Heckman (1976); the first of them provides the basis of Heckman's well-known two-step estimate to correct for selection bias. See § 6.2.7 for a development in this direction.

Another point of contact with our treatment arises from a variant of the above formulation, when the observed variable is $\min(Y, W)$, a case described by Maddala (2006) as an 'endogenous switching regression model'. In this situation, the two component equations of (3.40) play an equivalent role; consequently, to treat them on an equal footing, we do not assume that $\mathrm{var}\{\varepsilon_2\} = 1$. If $\mathrm{var}\{\varepsilon_2\} = \sigma^2$ and $\beta = \gamma$, then

$$
\min(Y, W) \sim \mathrm{SN}(x^\top \beta, \sigma^2, -\alpha(\delta)),
$$

recalling the argument leading to (2.16).

3.4.2 Stochastic frontier analysis

A theme of the econometric literature deals with the evaluation of efficiency of production units; this can be quantified as the output Q produced with a given amount of resources, or equivalently as the cost of resources required to produce a given amount of output. Important seminal papers are those of Aigner *et al.* (1977) and Meeusen and van den Broeck (1977). A more recent account is provided in chapters 9 and 10 of Coelli *et al.* (2005).

In the simplest formulation, a linear relationship is postulated between input and output variables after all variables have been transformed to the log scale. Hence, for a given production unit, its output on the log scale, say $Y = \log Q$, is written as

$$Y = x^\top \beta + V - U, \tag{3.41}$$

where x is a vector representing the log-transformed input factors employed to produce Y, β is an unknown vector of parameters, V and U are independent random components, with V symmetrically distributed around 0 and U positive.

Here $x^\top \beta$ represents the output produced by a technically efficient unit, V is a pure random error term and U is interpreted as the inefficiency of the given production unit. The term 'stochastic frontier model' for (3.41) expresses the idea that the frontier of technical efficiency $x^\top \beta$ may occasionally be exceeded, but this is only due to the purely erratic component V. In a more elaborate version of the model, $x^\top \beta$ is replaced by some non-linear function $h(x, \beta)$.

The dual model for production cost instead of output is similar to (3.41) but with the negative sign in front of U replaced by +, and of course the meanings of x and Y are changed. The two models can be expressed in the single form

$$\begin{aligned} Y &= x^\top \beta + V + s\,U \\ &= x^\top \beta + R\,, \end{aligned}$$

say, where $s = -1$ for the output model and $s = 1$ for the cost model.

Of the many possible options concerning U and V, a natural and indeed commonly adopted formulation is to assume that $V \sim \mathrm{N}(0, \sigma_v^2)$ and $U \sim \sigma_u \chi_1$. Taking into account the representation (2.14), it is immediate that

$$R = \omega \left(\sqrt{1 - \delta^2} V_0 + \delta |U_0| \right) \sim \mathrm{SN}(0, \omega^2, \alpha(\delta)),$$

where U_0 and $V_0 = \sigma_v^{-1} V$ are independent $N(0,1)$ variates, $|U_0| \stackrel{\mathrm{d}}{=} \sigma_u^{-1} U$ and

$$\omega^2 = \sigma_u^2 + \sigma_v^2, \qquad \delta = \frac{s\,\sigma_u}{(\sigma_u^2 + \sigma_v^2)^{1/2}}, \qquad \alpha(\delta) = s\,\frac{\sigma_u}{\sigma_v}.$$

In the specialized literature on stochastic frontier analysis, various parameterizations are employed, but the one in more common use, including the above-indicated accounts, appears to be $(\beta, \sigma^2, \lambda)$, as follows: $\beta = \beta^{\mathrm{DP}}$, $\sigma^2 = \omega^2$, $\lambda = |\alpha|$, which has an immediate correspondence with our DP set. In their context, the sign s does not need to be taken care of, since this is fixed at $s = -1$ or $s = 1$, depending on whether the output or the cost model is in use.

After a model of type (3.41) has been fitted to a set of data, it is of interest to produce, in addition to estimates of the parameters and other usual inferential summaries, an evaluation of the so-called technical efficiency of each production unit. In our notation, the problem amounts to evaluating the value taken on by U, for each nominated production unit or, on the natural scale of the observations, the value of $\exp(-U)$.

The problem involves consideration of the distribution of U conditional on the value of Y. If the parameters β are taken as known, this is the same as the conditional distribution of U for a fixed value of $R = V + sU = Y - x^\top \beta$. It is a simple exercise to obtain that, conditionally on $R = r$, the distribution of U is a truncated normal whose density function at u is

$$f_c(u|r) = \frac{1}{\sigma_c \Phi(\mu_c/\sigma_c)} \, \varphi\left(\frac{u - \mu_c}{\sigma_c}\right), \qquad u > 0, \qquad (3.42)$$

where

$$\mu_c = \frac{s\,r}{\sigma_v^2}\left(\frac{1}{\sigma_u^2} + \frac{1}{\sigma_v^2}\right)^{-1}, \qquad \sigma_c^2 = \left(\frac{1}{\sigma_u^2} + \frac{1}{\sigma_v^2}\right)^{-1} = \frac{\sigma_u^2 \sigma_v^2}{\sigma_v^2 + \sigma_u^2}.$$

The mean value of distribution (3.42) is computed by direct integration, which lends

$$\hat{u} = \mathbb{E}\{U|R = r\} = \mu_c + \sigma_c \frac{\varphi(\mu_c/\sigma_c)}{\Phi(\mu_c/\sigma_c)} = \mu_c + \sigma_c\,\zeta_1(\mu_c/\sigma_c),$$

and this expression can be used to estimate the technical efficiency via $\exp(-\hat{u})$, once one has obtained estimates of the parameters and the residual $r = y - x^\top \hat{\beta}$ for the nominated unit having $Y = y$. Notice that the observed value y affects μ_c, via r, but not σ_c.

Conversion of the inefficiency on the original scale is achieved in a simple form via $\exp(-\hat{u})$. To avoid bias due to the non-linear transformation, an alternative form of evaluation is provided by direct computation of

$$\mathbb{E}\{\exp(-U)|R = r\} = \int_0^\infty \exp(-u)\, f_c(u|r)\, du$$

$$= \exp\left(\tfrac{1}{2}\sigma_c^2 - \mu_c\right) \frac{\Phi(\mu_c/\sigma_c - \sigma_c)}{\Phi(\mu_c/\sigma_c)}.$$

3.5 Complements

Complement 3.1 (EM algorithm)   Given a random sample $y_1, \ldots, y_n$ from $SN(\xi, \omega^2, \alpha)$, we want to develop an EM algorithm to compute the MLE of the parameters.

One way to tackle the problem is based on the stochastic representation (2.13), which leads to consideration of the density function

$$f(u, v) = 2\,\varphi_B(u, v; \delta), \qquad (u, v) \in (\mathbb{R}^+ \times \mathbb{R}),$$

of a bivariate standard normal variable (U, V) truncated below $U = 0$; here φ_B denotes the bivariate standard normal density (B.14) when the correlation is $\delta = \delta(\alpha)$. The conditional density of U given $V = v$ is

$$f_c(u|v) = \frac{2\,f(u, v)}{2\,\varphi(v)\,\Phi(\alpha v)} = \frac{\varphi\left((u - \delta v)/\sqrt{1 - \delta^2}\right)}{\sqrt{1 - \delta^2}\,\Phi(\alpha v)}, \qquad u > 0,$$

such that

$$\hat{u}^{(1)} = \mathbb{E}\{U|v\} = \delta v + \zeta_1(\alpha v)\,\sqrt{1 - \delta^2},$$
$$\hat{u}^{(2)} = \mathbb{E}\{U^2|v\} = 1 - \delta^2 + (\delta v)^2 + \zeta_1(\alpha v)\,\delta\,\sqrt{1 - \delta^2}\, v,$$

where ζ_1 is given by (2.20).

An EM scheme can then be formulated regarding $\tilde{u}_i = \omega u_i$ as the missing observation of the pair $(\tilde{u}_i, y_i)$, where $y_i = \xi + \omega v_i$, for $i = 1, \ldots, n$. The contribution to the complete data log-likelihood from $(\tilde{u}_i, y_i)$ is

$$-\frac{1}{2}\log(1 - \delta^2) - 2\log\omega - \frac{(y_i - \xi)^2 - 2\delta\,(y_i - \xi)\,\tilde{u}_i + \tilde{u}_i^2}{2\,(1 - \delta^2)\,\omega^2},$$

where we regard δ as a parameter in place of α. The E-step of the EM algorithm is immediate from the above expressions of $\hat{u}^{(1)}$ and $\hat{u}^{(2)}$, replacing the parameter values by the estimates at the current iteration, leading to

$$\tilde{u}_i^{(1)} = \hat{\omega}\left(\hat{\delta}\,\hat{v}_i + \zeta_1(\hat{\alpha}\,\hat{v}_i)\,\sqrt{1 - \hat{\delta}^2}\right),$$
$$\tilde{u}_i^{(2)} = \hat{\omega}^2\left(1 - \hat{\delta}^2 + (\hat{\delta}\,\hat{v}_i)^2 + \zeta_1(\hat{\alpha}\hat{v}_i)\,\hat{\delta}\,\sqrt{1 - \hat{\delta}^2}\,\hat{v}_i\right),$$

where $\hat{v}_i = (y_i - \hat{\xi})/\hat{\omega}$. The M-step to get the new estimates is performed by solving

$$\hat{\xi} = \frac{\sum_i y_i - \hat{\delta} \sum_i \tilde{u}_i^{(1)}}{n}, \qquad \hat{\omega}^2 = \frac{Q(\hat{\xi})}{2n(1 - \hat{\delta}^2)}, \qquad \hat{\delta} = \frac{\sqrt{1 + 4r^2} - 1}{2r},$$

where

$$Q(\hat{\xi}) = \sum_i \left((y_i - \hat{\xi})^2 - 2\hat{\delta}(y_i - \hat{\xi})\tilde{u}_i^{(1)} + \tilde{u}_i^{(2)} \right), \qquad r = 2Q(\hat{\xi})^{-1} \sum_i (y_i - \hat{\xi})\tilde{u}_i^{(1)}.$$

The M-step involves an iterative procedure itself, but this can be accomplished very simply by repeated substitution.

   The following variant forms of the above algorithm are immediate: (a) the case with α fixed at a given value, (b) the case with both ω and α fixed; in these two cases, the M-step is non-iterative; (c) an extension to the regression case where $\xi_i = x_i^\top \beta$.

Problems

3.1 Consider a random sample $z_1, \ldots, z_n$ from $SN(0, 1, \alpha)$. We have seen that the MLE is infinite if all the z_i's have the same sign. When some of the z_i's are positive and some are negative, show that the likelihood equation for α has a finite root, and is unique (Martínez *et al.*, 2008). *Note:* This conclusion holds also for samples from any density of the form $2 f_0(x) \Phi(\alpha x)$.

3.2 Consider generalized forms of skew-normal densities formed by complementing (2.51) with location and scale parameters, hence of type

$$f(y) = \omega^{-1}\varphi(z) G_0(\alpha z), \qquad z = \omega^{-1}(y - \xi).$$

For estimating (ξ, ω, α), given a random sample from this distribution, prove that $(\bar{y}, s, 0)$ is still a solution of the likelihood equations, as in § 3.1.1, irrespective of G_0. Prove also that the observed information matrix is singular at point $(\bar{y}, s, 0)$ (Pewsey, 2006b).

3.3 Confirm the statement of the text that $a_k(\alpha)$ defined by (3.16) is equal to (3.33). Then show that $a_k(\alpha)$ is an even function or an odd function of α, depending on whether k is even or odd.

3.4 In Complement 3.1, we have seen a form of EM algorithm based on representation (2.13). Use the additive representation (2.14) to develop an alternative form of EM algorithm (Arellano-Valle *et al.*, 2005a; Lin *et al.*, 2007b).

3.5 Confirm the expressions of $\mathbb{E}\{U|v\}$ and $\mathbb{E}\{U^2|v\}$ in Complement 3.1.

4

Heavy and adaptive tails

4.1 Motivating remarks

The skew-normal density has very short tails. In fact, the rate of decay to 0 of the density $\varphi(x; \alpha)$ as $|x| \to \infty$ is either the same as the normal density or even faster, depending on whether x and α have equal or opposite sign, as specified by Proposition 2.8. This behaviour makes the skew-normal family unsuitable for a range of application areas where the distribution of the observed data is known to have heavier tails than the normal ones, sometimes appreciably heavier.

To construct a family of distributions of type (1.2) whose tails can be thicker than a normal ones, a solution cannot be sought by replacing the term $\Phi(\alpha x)$ in (2.1) with some other term $G_0\{w(x)\}$, since essentially the same behaviour of the SN tails would be reproduced. The only real alternative is to adopt a base density f_0 in (1.2) with heavier tails than the normal density.

For instance, we could select the Laplace density $\exp(-|x|)/2$, whose tails decrease at exponential rate, to play the role of base density and proceed along lines similar to the skew-normal case. This is a legitimate program, but it is preferable that f_0 itself is a member of a family of symmetric density functions, depending on a *tail weight parameter*, ν say, which allows us to regulate tail thickness. For instance, one such choice for f_0 is the Student's t family, where ν is represented by the degrees of freedom.

The idea of using families of densities symmetric about 0 and with adjustable tails, as a strategy to accommodate data distributions with long but otherwise unspecified tails, has been in use for quite some time. Specific formulations of this type have been put forward by Box and Tiao (1973, Section 3.2) and Lange *et al.* (1989), among others; additional similar work is quoted in these sources. The motivation of these formulations is to incorporate protection against the presence of outliers by adjusting the tail thickness according to the actual data behaviour. In this sense such a model provides a form of robust inference, although quite differently from the classical approach to robustness, linked to M-estimation and similar techniques.

A discussion on the relative merits of this approach to robustness and of the classical one will be given later in this chapter. For the moment we only recall that empirical studies, in particular Hill and Dixon (1982), indicate that often real data do not exhibit the extreme level of outlying observations employed in many theoretical robustness studies, while they often exhibit other forms of departure from normality much less examined. In particular, a frequent feature in real data is that outliers do not occur with the same frequency on both sides of the bulk of the distribution, that is, they are often placed asymmetrically in the two tails.

These empirical findings support the adoption of a parametric formulation which allows tail regulation combined with asymmetric behaviour. In our context, this translates into the combination of a symmetric family f_0, which allows tail regulation, with a factor $G_0\{w(x)\}$, which allows a different behaviour of the two tails. In this way we can produce a rich variety of shapes, allowing regulation of both skewness and kurtosis, and possibly even further features, depending on the complexity of $G_0\{w(x)\}$.

The next sections elaborate along the lines discussed above, for some choices of the class f_0. In all cases, the normal distribution is either included in the family f_0 or it is a limiting case, for a suitable sequence of v values.

4.2 Asymmetric Subbotin distribution

4.2.1 Subbotin distribution

Subbotin (1923) presented a parametric family of density functions on the real line, symmetric about 0 for all values of the positive parameter v, say, which regulates its tail weight. With slight modification of the original parameterization, the density function can be written as

$$f_v(x) = c_v \, \exp\left(-\frac{|x|^v}{v}\right), \qquad x \in \mathbb{R}, \tag{4.1}$$

where $c_v = \{2 \, v^{1/v} \, \Gamma(1 + 1/v)\}^{-1}$. Subsequent authors have denoted this distribution with a variety of names: exponential power distribution, generalized error distribution, normal of order v, generalized normal distribution, and possibly others. The parameterization in (4.1) is that of Vianelli (1963).

The key feature of this family is that we can regulate widely its tail thickness as v varies. This fact is illustrated in Figure 4.1 whose left-side panel displays the density function for some values $v \geq 2$; the right-side panel refers to $v \leq 2$. The value $v = 2$ corresponds to the standard normal distribution. Moreover, (4.1) includes as a special case the Laplace distribution if $v = 1$, and it converges pointwise to the uniform density on $(-1, 1)$ if $v \to \infty$.

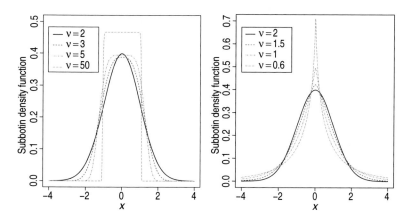

Figure 4.1 The Subbotin density functions when $v = 2, 3, 5, 50$ in the left-side panel, and $v = 2, 1.5, 1, 0.6$ in the right-side panel.

The possibility with (4.1) to produce both heavier and lighter tails than a normal distribution, depending on whether $v < 2$ or $v > 2$, is an interesting option. Although much emphasis is usually placed on the issue of heavy tails, the other case occurs too in real data. For instance, Cox (1977) reports that 'In a study I made some years ago of various kinds of routine laboratory tests in textiles, distributions with negative kurtosis occurred about as often as those with positive kurtosis'.

If Y is a random variable with density (4.1), then a standard computation indicates that $|Y|^v/v$ is a Gamma variable with shape parameter $1/v$ and scale factor 1. Reversing this relationship, a stochastic representation of Y is obtained as

$$Y = \begin{cases} (vX)^{1/v} & \text{with probability } 1/2, \\ -(vX)^{1/v} & \text{with probability } 1/2, \end{cases} \tag{4.2}$$

where $X \sim \text{Gamma}(1/v, 1)$. This provides a method to generate random numbers with density (4.1), since techniques for the generation of Gamma variates are commonly available. The same type of argument leads to the distribution function of (4.1), which is

$$F_\nu(x) = \frac{1}{2}\left\{1 + \text{sgn}(x)\frac{\gamma(|x|^\nu/\nu; 1/\nu)}{\Gamma(1/\nu)}\right\}, \qquad x \in \mathbb{R}, \qquad (4.3)$$

where $\gamma(u; \omega) = \int_0^u t^{\omega-1}e^{-t}\,dt$ denotes the incomplete Gamma function. Direct integration making use of the same variable transformation provides the mth moment

$$\mathbb{E}\{Y^m\} = \begin{cases} 0 & \text{if } m \text{ is odd,} \\ \dfrac{\nu^{m/\nu}\,\Gamma((m+1)/\nu)}{\Gamma(1/\nu)} & \text{if } m \text{ is even.} \end{cases} \qquad (4.4)$$

4.2.2 Asymmetric versions of Subbotin distribution

Given the remarks of Section 4.1, we considered asymmetric versions of the Subbotin distribution, proceeding similarly to the skew-normal construction. Replacing φ and Φ in (2.3) by f_ν and $G_0(\cdot)$, respectively, we arrive at the density function

$$f(x) = \frac{2}{\omega}\,f_\nu\left(\frac{x-\xi}{\omega}\right)\,G_0\left(\alpha\frac{x-\xi}{\omega}\right), \qquad x \in \mathbb{R}, \qquad (4.5)$$

where G_0 is as required in Proposition 1.1, and ξ and ω are location and scale parameters ($\omega > 0$).

Of the many options available for G_0, two have received attention, called type I and type II. Type I will be summarized in Complement 4.1. The rest of this section deals with the asymmetric Subbotin distribution of type II, AS2 for short, which is produced when $G_0(\cdot)$ in (4.5) is taken equal to

$$G_0(t) = \Phi\left(\text{sgn}(t)\frac{|t|^{\nu/2}}{\sqrt{\nu/2}}\right), \qquad t \in \mathbb{R}, \qquad (4.6)$$

which corresponds to the distribution function of $\text{sgn}(U)\left|U\sqrt{\nu/2}\right|^{2/\nu}$ when $U \sim \text{N}(0, 1)$. If $\nu = 2$, AS2 reduces to the SN distribution. Figure 4.2 displays the resulting density function for some values of ν and α.

If Z denotes a random variable with this distribution when $\xi = 0$ and $\omega = 1$, $\mathbb{E}\{Z^m\}$ with m even is computed from (4.4). If m is odd, we make use of (1.20) where in this case V has density function $2\,f_\nu(x)$ for $x > 0$ and 0 otherwise. For $\alpha > 0$, let $s = (m+1)/\nu$ and write

$$\mathbb{E}\{V^m\,G_0(\alpha V)\} = 2c_\nu \int_0^\infty x^m \exp\left(-\frac{x^\nu}{\nu}\right)\Phi\left(\frac{(\alpha x)^{\nu/2}}{\sqrt{\nu/2}}\right)dx$$

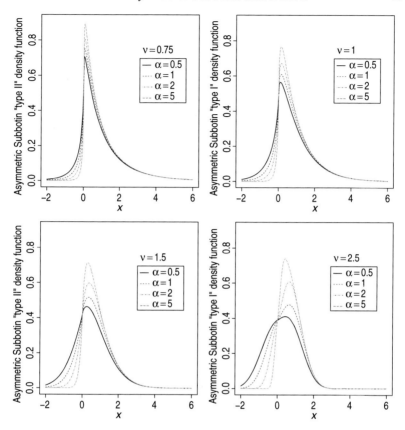

Figure 4.2 Asymmetric version II of Subbotin distribution when $\xi = 0$, $\omega = 1$, $\alpha = 0.5, 1, 2, 5$ and $\nu = 0.75, 1, 1.5, 2.5$.

$$= \frac{\nu^{m/\nu}}{\Gamma(1/\nu)} \int_0^\infty e^{-u} u^{s-1} \Phi\left(\sqrt{2\alpha^\nu u}\right) du$$

$$= \frac{\nu^{m/\nu}\,\Gamma(s)}{\Gamma(1/\nu)}\, \mathbb{E}\left\{\sqrt{2\,s\alpha^\nu W}\right\}$$

$$= \frac{\nu^{m/\nu}\,\Gamma(s)}{\Gamma(1/\nu)}\, T\left(\sqrt{2\,s\,\alpha^\nu};\, 2s\right),$$

where $W \sim \chi^2_{2s}/(2s)$, $T(t;\rho)$ denotes the Student's distribution function on ρ d.f., and the last equality is based on (B.12) on p. 233. Combining these elements, the general expression of $\mathbb{E}\{Z^m\}$ is

$$\mathbb{E}\{Z^m\} = \frac{\nu^{m/\nu}\,\Gamma(s)}{\Gamma(1/\nu)} \times \begin{cases} 1 & \text{if } m \text{ is even,} \\ \mathrm{sgn}(\alpha)\, Q & \text{if } m \text{ is odd,} \end{cases} \qquad (4.7)$$

where, on setting $\Delta = |\alpha|^\nu/(1 + |\alpha|^\nu)$,

$$
\begin{aligned}
Q &= 2\,T\left(\sqrt{2\,s\,|\alpha|^\nu};\, 2s\right) - 1 \\
&= I_\Delta(\tfrac{1}{2}, s) \\
&= \Delta^{s-1/2} \sum_{j=0}^{s-1} \binom{s - \tfrac{1}{2}}{j} |\alpha|^{-js} \qquad \text{if } s \text{ is an integer,}
\end{aligned}
$$

taking into account the relationship between the t and the Beta distribution functions, and known properties of the incomplete Beta function $I_x(a, b)$.

It can be shown by direct computation of the second derivative of the log-density that this is concave for $\nu > 1$.

With the aid of (4.7), we can produce Figure 4.3 which displays in a graphical form the range of coefficients of skewness and kurtosis, γ_1 and γ_2, for a few values of ν, as α ranges on the positive semi-axis; if α is negative, the curves are mirrored on the opposite side of the vertical axis. The curve with $\nu = 2$ corresponds to the one in Figure 2.2. Notice the curl of the curves with ν just larger than 2, which causes them to intersect. This implies that some members of this family with $\nu > 2$ cannot be separated on the basis of the first four moments only. For $\nu < 2$, the observed ranges of γ_1 and γ_2 are quite wide.

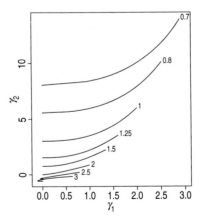

Figure 4.3 Asymmetric version of Subbotin distribution: each curve denotes the points (γ_1, γ_2) for the type II version as α varies along the positive half-line, for a fixed value of ν indicated next to the curve.

Bibliographic notes

Azzalini (1986) has studied two asymmetric forms of the Subbotin distribution (denoted type I and type II, corresponding to AS1 and AS2 here) as a means of rendering likelihood inference more robust to the presence of outliers, allowing explicitly for different placement in the tails. The AS1 variant as been employed by Cappuccio *et al.* (2004) as the error term distribution of a stochastic volatility time series model. DiCiccio and Monti (2004) have developed asymptotic theory results for maximum likelihood estimates of the AS2 parameters, examining in detail the case of $\alpha = 0$.

4.3 Skew-*t* distribution

4.3.1 Definition and main properties

Another set of densities symmetric about 0 and commonly employed when we need to regulate the tail thickness is the Student's *t* family. Its density function is

$$t(x; v) = \frac{\Gamma(\frac{1}{2}(v+1))}{\sqrt{\pi v}\,\Gamma(\frac{1}{2}v)}\left(1 + \frac{x^2}{v}\right)^{-(v+1)/2}, \qquad x \in \mathbb{R}, \qquad (4.8)$$

where $v > 0$ denotes the degrees of freedom (d.f.). Recall that the tails of this density are always heavier than those of the normal one, for finite v. Therefore, using the mechanism of Proposition 1.1, we can produce distributions with at most one tail lighter than the normal, while the other tail will always be heavier for any finite v. By analogy with various cases examined so far, it would be instinctive to introduce an asymmetric version of this density in the linear form

$$2\,t(x; v)\,T(\alpha x; v), \qquad (4.9)$$

where $T(.; v)$ denotes the distribution function of (4.8), and α a slant parameter. Although this route is legitimate, there are reasons for preferring another type of construction.

One of these reasons is to maintain a similarity with the familiar construction of a *t* variate, that is, the representation

$$Z = \frac{Z_0}{\sqrt{V}}, \qquad (4.10)$$

where $Z_0 \sim N(0, 1)$ and $V \sim \chi_v^2/v$ are independent variates.

Correspondingly, a reasonable formulation of an asymmetric *t* distribution is obtained by replacing the assumption on the distribution of Z_0 with

$Z_0 \sim SN(0, 1, \alpha)$. In this case, if $h(\cdot)$ denotes the density function of V, that of Z is

$$
\begin{aligned}
t(x; \alpha, \nu) &= \int_0^\infty 2\, \varphi(x\sqrt{t})\, \Phi(\alpha x \sqrt{t})\, \sqrt{t}\, h(t)\, dt \\
&= \frac{2}{\Gamma(\frac{1}{2}\nu)\sqrt{\pi\nu}} \left(1 + \frac{x^2}{\nu}\right)^{-(\nu+1)/2} \int_0^\infty e^{-u} u^{(\nu-1)/2} \Phi\left(\frac{\alpha x \sqrt{2u}}{\sqrt{x^2+\nu}}\right) du \\
&= 2\, t(x; \nu)\, T\left(\alpha x \sqrt{\frac{\nu+1}{\nu+x^2}}; \nu+1\right),
\end{aligned}
\tag{4.11}
$$

where the last equality makes use of (B.12) on p. 233. If $\alpha = 0$, (4.11) reduces to the usual Student's t density. If $\nu \to \infty$, (4.11) converges to the $SN(0, 1, \alpha)$ density, as expected from representation (4.10).

When Z_0 in (4.10) is $N(0, 1)$, the resulting Student's t distribution can be viewed as a scale mixture of normals. Similarly now, with $Z_0 \sim SN(0, 1, \alpha)$, the resulting distribution is a scale mixture of SN variables.

If we write the final factor of (4.11) as $G_0\{w(x)\} = T(w(x); \nu + 1)$, then G_0 is not the integral of the base density $t(x; \nu)$ and

$$
w(x) = \alpha x \sqrt{\frac{\nu+1}{\nu+x^2}}
\tag{4.12}
$$

is clearly non-linear, at variance with αx in (4.9); the departure from linearity is appreciable if ν is small. However, $w(x)$ is odd and (4.11) still falls within the pattern of (1.2). From the property of modulation invariance, we can state that

$$
Z^2 \sim F(1, \nu),
\tag{4.13}
$$

where $F(\nu_1, \nu_2)$ denotes the Snedecor's F distribution with ν_1 and ν_2 d.f.

In addition to representation (4.10), there are other reasons to prefer (4.11) over the form (4.9) with linear $w(x)$. One of these is the simpler mathematical treatment, for instance for computing moments. Additional important reasons will appear when we discuss the multivariate version of the distribution.

Figure 4.4 displays the graphical appearance of (4.11) for two values of α and a selected range of ν. As usual, only positive α's need to be considered. Since the regular t density (4.8) is not log-concave, inevitably this property cannot hold for (4.11). However, the symmetric t density is unimodal, and it can be proved that the same holds true for the skew-t density as well; see Problem 4.7.

For applied work, we need to extend the family (4.11) to include a location and a scale parameter. Similarly to (2.2), consider $Y = \xi + \omega Z$, leading

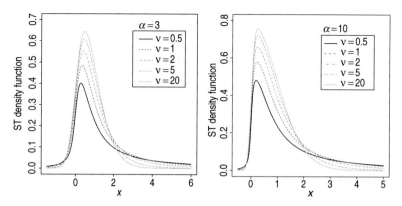

Figure 4.4 Skew-*t* density functions when $\alpha = 3$ in the left-side panel and $\alpha = 10$ in the right-side panel, for a few values of *v*.

to a four-parameter family of distributions whose density function at x is $\omega^{-1} t(z; \alpha, \nu)$, where $z = \omega^{-1}(x - \xi)$. We shall say that Y has a skew-*t* (ST) distribution and write

$$Y \sim \mathrm{ST}(\xi, \omega^2, \alpha, \nu).$$

From representation (4.10), the mth moment of Z is simply expressed as

$$\mathbb{E}\{Z^m\} = \mathbb{E}\{V^{-m/2}\} \, \mathbb{E}\{Z_0^m\},$$

whose components are provided by the standard result

$$\mathbb{E}\{V^{-m/2}\} = \frac{(\nu/2)^{m/2} \, \Gamma\left(\tfrac{1}{2}(\nu - m)\right)}{\Gamma\left(\tfrac{1}{2}\nu\right)}, \qquad \text{if } m < \nu, \qquad (4.14)$$

and by the expressions of $\mathbb{E}\{Z_0^m\}$ given in § 2.1.4. On denoting

$$b_\nu = \frac{\sqrt{\nu}\, \Gamma\left(\tfrac{1}{2}(\nu - 1)\right)}{\sqrt{\pi}\, \Gamma\left(\tfrac{1}{2}\nu\right)}, \qquad \text{if } \nu > 1, \qquad (4.15)$$

some simple algebra gives

$$\mu = \mathbb{E}\{Y\} = \xi + \omega \, b_\nu \, \delta, \qquad\qquad\qquad\qquad \text{if } \nu > 1, \qquad (4.16)$$

$$\sigma^2 = \mathrm{var}\{Y\} = \omega^2 \left[\frac{\nu}{\nu - 2} - (b_\nu \, \delta)^2\right] = \omega^2 \sigma_z^2, \text{ say}, \qquad \text{if } \nu > 2, \qquad (4.17)$$

$$\gamma_1 = \frac{b_\nu \, \delta}{\sigma_z^{3/2}} \left[\frac{\nu(3 - \delta^2)}{\nu - 3} - \frac{3\,\nu}{\nu - 2} + 2\,(b_\nu \, \delta)^2\right], \qquad\qquad \text{if } \nu > 3, \qquad (4.18)$$

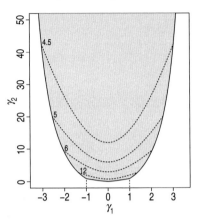

Figure 4.5 Skew-t distribution: the grey area denotes the lower portion of the admissible region of (γ_1, γ_2) if $v > 4$. The dashed lines labelled $4.5, 5, 6, 12$ denote the curves associated with that value of v when α spans the real line.

$$\gamma_2 = \frac{1}{\sigma_z^4}\left[\frac{3v^2}{(v-2)(v-4)} - \frac{4(b_v\delta)^2 v(3-\delta^2)}{v-3} + \frac{6(b_v\delta)^2 v}{v-2} - 3(b_v\delta)^4\right] - 3$$

$$\text{if } v > 4, \qquad (4.19)$$

where δ is as in (2.6), and γ_1 and γ_2 represent the third and fourth standardized cumulants of Y, respectively.

The shaded area in Figure 4.5 displays the lower portion of the admissible region of (γ_1, γ_2) when $v > 4$. The actual range of the coefficient of excess kurtosis γ_2 goes up to ∞ as $v \to 4$. The dashed lines inside this region correspond to a few specific choices of v. The portion of the lower boundary of the shaded region where $0 \le \gamma_1 < 0.995\ldots$ coincides with the curve in Figure 2.2 for the SN distribution. The range of the coefficient of skewness γ_1 is $(-4, 4)$ if $v > 4$, but it becomes the whole real line if we consider $v > 3$.

Bibliographic notes

The above discussion summarizes some factors extracted from a more general development that was originally framed in a multivariate setting, and will be discussed in Chapter 6. This route has been adopted to ease exposition. A construction which includes the multivariate version of the ST distribution has been presented by Branco and Dey (2001). Their expression

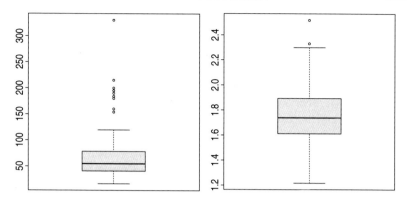

Figure 4.6 Price of a bottle of Barolo: boxplot of the value in euros on the left side; boxplot of the $\log_{10}$-transformed data on the right side.

of the ST density was, however, more implicit than (4.11), which was obtained independently Gupta (2003) and Azzalini and Capitanio (2003); the latter paper includes various other results on the ST distribution reported in earlier pages. Much additional work has appeared in the literature along this line, some of which will be mentioned later on in connection with statistical aspects.

In parallel, a range of alternative proposals in the literature have adopted the term 'skew-*t* distribution' referring to mathematically different constructions, even if of broadly similar motivation. Furthermore, the classical Pearson type-IV distribution is sometimes called 'skewed-*t*'. To avoid confusion, the reader should bear in mind this ambiguity in the terminology.

4.3.2 Remarks on statistical aspects

It is well known that the price of a bottle of wine can vary widely. This is confirmed once more by the data represented graphically as a boxplot in the left panel of Figure 4.6, which refers to 103 quotations for bottles of Barolo wine produced by a number of wineries in the Piedmont region of Italy. The prices, in euros, have been recorded from the website of a single reseller in July 2010, and they all refer to a standard size bottle (75 cl). Prices are influenced by a variety of factors, such as age, which in this case ranges from 4 to 33 years. However, for simplicity, we examine the price distribution unconditionally.

A standard device to handle marked skewness of data is to log-transform them. For these data, the outcome using base-10 logarithms is displayed in the right panel of Figure 4.6. Although much diminished, skewness is still present in the new boxplot, and departure from normality can be further confirmed by the Shapiro–Wilk test whose observed significance level is about 0.6%. We then proceed by fitting the distributions described in the previous sections to the log-transformed prices. Although it would then make sense to fit these distributions to the original data, the strong asymmetry of the data would make the fitting process less manageable. Also, the adopted route is more convenient for our illustrative purposes.

Figure 4.7 displays the histogram of the log-transformed data with four fitted densities superimposed. Three of these curves belong to the parametric families described earlier in this chapter, that is the two asymmetric forms of Subbotin distributions and the skew-*t* distribution. The fourth curve, which derives from an entirely different construction type, is an alternative form of asymmetric *t* distribution introduced by Jones (2001) and studied in detail by Jones and Faddy (2003); its density function for zero location and unit scale parameter is

$$\text{constant} \times \left(1 + \frac{x}{\sqrt{a + b + x^2}}\right)^{a+1/2} \left(1 - \frac{x}{\sqrt{a + b + x^2}}\right)^{b+1/2}, \qquad x \in \mathbb{R},$$

(4.20)

where a and b are positive parameters. For all the above families, the parameters have been estimated by maximum likelihood.

It is apparent from Figure 4.7 that all these parametric families fit the observed data distribution quite well, and the four curves are very similar to each other. The close similarity of the fitted distributions is not episodic, but observed in many other cases. The near-equivalence of these parametric families in terms of fitting adequacy can be viewed as an effect of their flexibility. The essential equivalence of the fitted distribution is especially true for the skew-*t* forms (4.11) and (4.20), which share a similar tail behaviour of polynomial type, originating from their Student's *t* imprint.

The implication of these remarks is that, in many practical cases, the choice among alternative parametric families must be largely based on criteria other than their ability to fit data. For instance, a simple mathematical expression of the density is appreciated, and (4.20) is attractive in this sense. An important aspect to take into account is the existence of a multivariate version of the chosen parametric family. A discussion of multivariate distributions will take place later in this book, but it is an advantage that a multivariate version of a given distribution exists. It is quite often the case

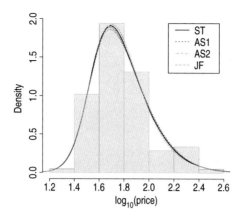

Figure 4.7 Price of a bottle of Barolo: histogram of the $\log_{10}$-transformed data and four fitted distributions, denoted as follows. ST: skew-*t* distribution; AS1: asymmetric Subbotin distribution of type I; AS2: asymmetric Subbotin distribution of type II, JF: Jones and Faddy distribution.

that a set of univariate data analyses is followed by a joint modelling of the same variables, and it is then preferable to retain the same type of parametric family moving from the univariate to the multivariate context. In this sense the Subbotin-type distributions are not really satisfactory: while a multivariate version of the symmetric Subbotin distribution exists, and so the same holds also for its asymmetric versions, these parametric classes are not closed under marginalization; in other words, the marginal components of the multivariate Subbotin distribution are not members of the same class (Kano, 1994), and the same fact carries over to asymmetric versions. The ST distribution (4.11) is then preferable on this front, since a multivariate version exists, and it is closed under marginalization. For distribution (4.20), a multivariate version is currently available in a form with restricted support; see Jones and Larsen (2004).

Another relevant aspect is the regularity of the log-likelihood function, or equivalently of the deviance function. The left-hand panel of Figure 4.8 displays the profile deviance function for the parameter α of the two versions of the asymmetric Subbotin distribution and for the ST family. Formally, the meaning of α in these three families is not identical, but the analogy in their construction mechanism makes the deviance functions

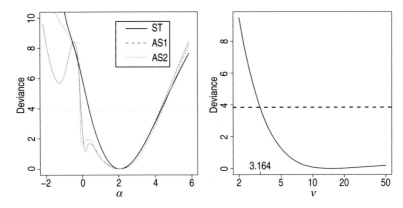

Figure 4.8 Log-price of a bottle of Barolo: profile deviance function for α for distributions AS1, AS2 and ST (left panel) and for v for ST only, plotted on the log-scale (right panel).

broadly comparable. In fact, the MLE of α is nearly the same for all of them, and so is the behaviour of the deviance for α larger than about 2. On the contrary, below 2 the three curves diverge substantially: while the ST deviance is smooth and monotonic, the other two are both non-monotonic and fairly irregular. This rather unpleasant behaviour is not systematic but not infrequent with the asymmetric Subbotin distributions. Jones and Faddy (2003) propose that the parameter representing asymmetry of (4.20) is taken to be $q = (a - b)/\sqrt{ab\,(a + b)}$; for the data under consideration, its estimate is $\hat{q} = 0.26$. Since q is completely unrelated to α, a direct graphical comparison of the deviance function with the other curves in the left panel of Figure 4.8 is not feasible. However its plot, not reproduced here, has a perfectly smooth behaviour, qualitatively similar to that for the ST distribution.

Although there are arguments in favour and against each of the distributions considered, the above discussion points towards the ST family as a convenient general-purpose choice. Obviously, a statement of this kind cannot be taken as a universal rule; specifically, the ST family is not suitable for handling situations where both tails are shorter than the normal ones. However, with this *caveat*, the ST family is the one we shall focus on.

4.3.3 The log-likelihood function and related quantities

It will have been noticed that none of the three profile deviance curves in the left panel of Figure 4.8 has a stationary point at $\alpha = 0$, and therefore the same holds true for the corresponding log-likelihood functions. There is then a marked difference from the analogous functions of the SN distribution, which always produces a stationary point at $\alpha = 0$. In Chapter 3, this fact implied important consequences for likelihood inference when the parameter set is in a neighbourhood of $\alpha = 0$. This is one of the points we want to explore in the following. As anticipated, we now focus on the ST distribution.

The contribution from a single observation y to the ST log-likelihood of the direct parameters $\theta^{\mathrm{DP}} = (\xi, \omega, \alpha, \nu)$ is

$$\ell_1(\theta^{\mathrm{DP}}; y) = \mathrm{constant} - \log \omega - \tfrac{1}{2}\log \nu + \log \Gamma(\tfrac{1}{2}(\nu + 1)) - \log \Gamma(\tfrac{1}{2}\nu)$$
$$- \tfrac{1}{2}(\nu + 1)\log\left(1 + \frac{z^2}{\nu}\right) + \log T(w; \nu + 1), \qquad (4.21)$$

where

$$z = \frac{y - \xi}{\omega}, \qquad w = w(z) = \alpha z r, \qquad r = r(z, \nu) = \sqrt{\frac{\nu + 1}{\nu + z^2}}.$$

Differentiation followed by some algebraic reduction gives the components of the score vector as

$$\frac{\partial \ell_1}{\partial \xi} = \frac{z r^2}{\omega} - \frac{\alpha \nu\, r\, h(w)}{\omega(\nu + z^2)},$$

$$\frac{\partial \ell_1}{\partial \omega} = -\frac{1}{\omega} + \frac{(z r)^2}{\omega} - \frac{\nu\, w\, h(w)}{\omega(\nu + z^2)},$$

$$\frac{\partial \ell_1}{\partial \alpha} = z r\, h(w), \qquad (4.22)$$

$$\frac{\partial \ell_1}{\partial \nu} = \frac{1}{2}\left[\psi(\tfrac{1}{2}\nu + 1) - \psi(\tfrac{1}{2}\nu) - \frac{2\nu + 1}{\nu(\nu + 1)} - \log\left(1 + \frac{z^2}{\nu}\right)\right.$$
$$\left. + \frac{(z r)^2}{\nu} + \frac{\alpha z(z^2 - 1)\, h(w)}{(\nu + z^2)^2\, r} + \frac{g(\nu)}{T(w; \nu + 1)}\right],$$

where $\psi(x) = \mathrm{d}\log \Gamma(x)/\mathrm{d}x$ is the digamma function and

$$h(w) = \frac{t(w; \nu + 1)}{T(w; \nu + 1)},$$

$$g(\nu) = \frac{\mathrm{d}\,T(w(\alpha z\, r(z, \nu)); \nu + 1)}{\mathrm{d}\nu}.$$

$$= \int_{-\infty}^{w} \left[\frac{(v+2) x^2}{(v+1)(v+1+x^2)} - \log\left(1 + \frac{x^2}{v+1}\right) \right] t(x; v+1) \, dx.$$

In a regression formulation analogous to (3.8), the first component of the score function (4.22) is replaced by

$$\frac{\partial \ell_1}{\partial \beta} = \left(\frac{z \, r^2}{\omega} - \frac{\alpha v \, r \, h(w)}{\omega \, (v+z^2)} \right) x. \qquad (4.23)$$

When v is regarded as fixed, from (4.22) it is easy to see why the ST log-likelihood function at $\alpha = 0$ behaves differently from the SN case. The origin of the anomalies encountered in the SN case lies in the proportionality of the first and third components of the score function (3.2) for any fixed parameter set $(\xi, \omega, 0)$, when viewed as a function of the sample values. The same proportionality does not hold for similar components of the ST score function (4.22), which at $\alpha = 0$ become

$$u_0(\xi) = \left. \frac{\partial \ell_1}{\partial \xi} \right|_{\alpha=0} = \frac{z}{\omega} \frac{v+1}{v+z^2},$$

$$u_0(\alpha) = \left. \frac{\partial \ell_1}{\partial \alpha} \right|_{\alpha=0} = h(0) \, z \sqrt{\frac{v+1}{v+z^2}},$$

where $h(0) = 2 \, t(0; v+1)$ is a constant. Since $u_0(\alpha)/u_0(\xi) \propto \sqrt{v+z^2}$, then $u_0(\xi)$ and $u_0(\alpha)$, viewed as functions of z, are non-proportional. This implies that the joint distribution of their underlying random variables is not degenerate and the phenomenon of rank-deficiency of the expected Fisher information at $\alpha = 0$ does not occur.

Moreover, when a random sample $y_1, \ldots, y_n$ is available and we add up the terms $u_0(\xi)$ and $u_0(\alpha)$ evaluated at $z_i = (y_i - \xi)/\omega$, for $i = 1, \ldots, n$, any pair $(\tilde{\xi}, \tilde{\omega})$ which solves the first two likelihood equations, that is, equates to 0 both the sum of terms $u_0(\xi)$ and the sum of similar terms $u_0(\omega)$, does not in general also equate to 0 the sum of terms $u_0(\alpha)$. Hence the derivative of the profile log-likelihood, analogous to (3.7), does not vanish systematically at $\alpha = 0$.

In the four-parameter case, where v is estimated as well, it is more difficult to follow the same type of argument, because the component $\partial \ell_1 / \partial v$ of the score function (4.22) has a much more involved expression. However, numerical inspection of the profile log-likelihood has never indicated a stationary point, in any practical case considered.

The lack of a stationary point of the deviance function using the ST distribution is illustrated further by the left panel of Figure 4.9, which displays the deviance profile of the pair $(\alpha, \log v)$ for the log-price data of

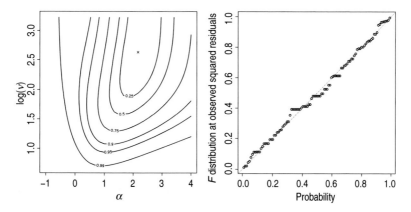

Figure 4.9 Log-price of a bottle of Barolo: profile deviance function for $(\alpha, \log \nu)$ for ST (left panel) and PP-plot of residuals (right panel).

Figure 4.6(b). The contour lines exhibit a smooth regular behaviour, delimiting convex regions, without any kink like in Figure 3.3(b). The levels of these curves are chosen equal to the percentage points of the χ_2^2 distribution, at the levels indicated by the labels, so that the enclosed regions represent confidence regions at the approximated confidence levels indicated.

For the same data, the right panel of Figure 4.9 illustrates the use of property (4.13) to build a diagnostics plot for residuals $\hat{z}_i$ of type (3.12) on p. 61 in a similar fashion to the PP-plot in Figure 3.7(b), now using the $F(1, \nu)$ distribution as reference in place of χ_1^2. In practice, ν and other parameters are replaced by estimates. A QQ-plot diagnostic, analogous to Figure 3.2, could equally be considered.

The penalized log-likelihood presented in § 3.1.8 to avoid boundary estimates for the SN family can be extended to the ST family. Even in this case a linear interpolation similar to (3.34) works well, with the only difference being that the coefficients, $e_{1\nu}$ and $e_{2\nu}$ say, now vary with ν. Similarly to the SN case, $e_{1\nu}$ has a simple expression while $e_{2\nu}$ requires numerical evaluation. An additional interpolation allows us to approximate $e_{2\nu}$ closely, for any given ν, without having to compute it numerically for each given value of ν.

Bibliographic notes

Initial work using the ST distribution for statistical analysis using regression models has been done by Branco and Dey (2002) in the Bayesian framework and by Azzalini and Capitanio (2003) in the classical approach;

the latter paper deals also with the multivariate case. The issue of lack of stationarity of the log-likelihood function at $\alpha = 0$, which had been observed numerically earlier, has been studied by Azzalini and Genton (2008), in connection with the use of the ST distribution as a tool for robust inference, which we shall discuss later.

Expressions (4.22), which are specific to the univariate ST distribution and avoid an approximation of earlier existing expressions, are as given by DiCiccio and Monti (2011), up to a change of notation. Additional results of this paper include, among others, the observed information matrix and the asymptotic distribution of the MLE when $\nu \to \infty$, tackled via the re-parameterization $\kappa = \nu^{-1}$. Since $\kappa = 0$ is a boundary point of the parameter space, a non-standard asymptotic distribution arises.

Details for the application of the MPLE method to the ST case are given by Azzalini and Arellano-Valle (2013). Related work on this problem has been done by Lagos Álvarez and Jiménez Gamero (2012).

4.3.4 *Centred parameters and other summary quantities*

The lack of singularity of the information matrix at $\alpha = 0$ eliminates the original motivation for the introduction of the CP parameterization in the SN case. However, the fact remains that the DP quantities are not so easily interpretable as the corresponding CP quantities, which are more familiar. For this reason, even for the ST distribution we may still wish to adopt the CP as summary quantities of a fitted model. The CP components are now $(\mu, \sigma, \gamma_1, \gamma_2)$, given by (4.16)–(4.19). In the regression case, the adjustment analogous to (3.23) on p. 67 is

$$\beta_0^{\mathrm{CP}} = \beta_0^{\mathrm{DP}} + \omega\, b_\nu\, \delta, \qquad (4.24)$$

provided $\nu > 1$.

We illustrate the question with the aid of Figure 4.10, which refers to the content of phosphate (x) and magnesium (y) in the Grignolino wine data. Similarly to the data of Figure 3.7, we fit a simple regression line of type (3.27), where now the assumption for the error term is $\varepsilon \sim \mathrm{ST}(0, \omega^2, \alpha, \nu)$. The left panel displays the scatter plot of the data, with the line fitted by MLE to the model just described and two additional lines superimposed: the least-squares fit and the line estimated by the robust MM method of Yohai (1987).

Similarly to Figure 3.7, the line $\hat{\beta}_0 + \hat{\beta}_1 x$ of the ST model lies too low in the cloud of points and an adjustment of the intercept is required. The right panel of Figure 4.10 displays, together with the MM and the earlier

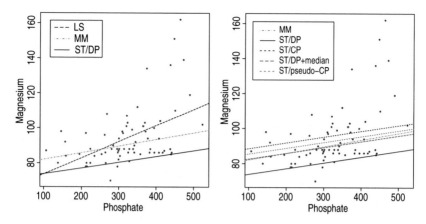

Figure 4.10 Wines data, Grignolino cultivar: magnesium versus phosphate content with superimposed regression lines. Left panel: least-squares line (LS), robust method fit (MM), linear model with ST errors in DP parameterization. Right panel: robust method and ST fit, the latter with various adjustments of the intercept as described in the text.

ST fit, other lines which differ from the latter only by a different intercept. The direct adaptation of the adjustment considered for Figure 3.7 is to replace $\hat{\beta}_0$ by $\hat{\beta}_0 + \hat{\mathbb{E}}\{\varepsilon\}$. However, the plot makes clear that the correction is excessive. The reason is that in this case there is a pronounced asymmetry ($\hat{\alpha} = 5.31$) and a very long tail, since $\hat{\nu} = 2.06$. Consequently, even if $\mathbb{E}\{\varepsilon\}$ exists, this type of correction tends to give quite high values. In fact, the correction diverges when $\hat{\nu}$ approaches 1 and becomes undefined when $\hat{\nu} \leq 1$. With $\hat{\nu} = 2.06$, the σ component of CP would still be computable, but not the other two components.

To overcome the instability or possibly the non-existence of CP, two routes are as follows. One direction is to make use of suitable quantile-based measures instead of moment-based measures. For the regression setting just discussed, this means adjusting the intercept by the median of ε instead of its mean value. The corresponding line for the above example is shown in the right panel of Figure 4.10 to be close to the MM line. Other quantile-based quantities could be introduced for the other components. For instance, we could use the semi-interquartile difference to measure dispersion, and other quantile-based measures of skewness and kurtosis.

An alternative route to cope with non-existence of moments is to introduce some form of shrinkage which prevents them from diverging. For instance, we can compute a shrunk form of mean by using (4.16) with ν

incremented by 1, (4.17) with v incremented by 2, and so on for (4.18) and (4.19). The resulting quantities $(\tilde{\mu}, \tilde{\sigma}, \tilde{\gamma}_1, \tilde{\gamma}_2)$ are called pseudo-CP. An advantage of this scheme over the use of quantile-based measures is to handle smoothly the transition from the ST to the SN case when $v \to \infty$, since the shrinkage effect then vanishes and we recover the CP for SN. The right panel of Figure 4.10 displays the effect of incrementing $\hat{\beta}_0$ by the pseudo-mean $\tilde{\mu}$ of the estimated distribution of ε.

Bibliographic notes and technical details

The observed DP information matrix $\mathcal{J}(\hat{\theta}^{\mathrm{DP}})$ can be converted into that of $\hat{\theta}^{\mathrm{CP}}$ similarly to (3.24); the inverse of the required Jacobian matrix D is given in Appendix B of Arellano-Valle and Azzalini (2013). From here we can obtain standard errors of the MLEs. Appendix A of the same paper tackles the question of whether the transformation from DP to CP is invertible, i.e., whether the CP set provides a proper parameterization of the ST family. The answer is essentially positive, although one stage of the argument involves a numerical optimization step, outside the formal rules of mathematical proof. Much of the rest of this paper is dedicated to a discussion of the pseudo-CP idea and its extension to the multivariate case, which requires substantial algebraic work. The pseudo-CP set is not an invertible transformation of the DP set of parameters.

4.3.5 Adaptive tails and robustness

In the numerical example of § 4.3.4, illustrated graphically in Figure 4.10, the data fit obtained under the ST assumption for the error terms was close to that produced by the MM-estimation method, which is credited with high robustness and high efficiency properties.

The natural question then is whether this observed closeness in that example was merely an accident or a regular fact. To gain more numerical evidence, let us consider another example. Figure 4.11 displays the phone-calls data employed by Yohai (1987) to illustrate the newly presented MM-estimates. Here the horizontal axis represents the sequence of years 1950 to 1973, say $1900 + x$, and the vertical axis represents the international phone-calls, y, made from Belgium in that period. The peculiar aspect of these data is that the response variable, y, was recorded in a non-homogeneous form over the years: for most cases, y represents the number of calls (denoted by 'N'), in tens of millions, but for the period 1964 to 1969 y represents the overall minutes of conversation (denoted by 'T'), in millions.

A simple linear regression has been fitted to the data by least squares,

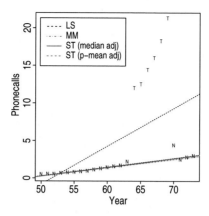

Figure 4.11 Phone-call data of Yohai (1987): the left plot
displays the scatter of data (x, y), where x represents years past
1900 and y represents the international phone-calls from Belgium,
measured either as number of calls (labelled N) or overall time
(labelled T), and four fitted lines as described in the text.

leading to the estimated line $-26.03 + 0.505\,x$. When superimposed on
the scatter plot in Figure 4.11, this line lies in an intermediate position
between the T and the N points. On the contrary, the MM line, which is
$-5.24 + 0.110\,x$, interpolates the majority group of points, that is the N's,
essentially discarding the T's.

We also consider a regression model with ST errors. The MLE of the DP
parameters $(\beta_0, \beta_1, \omega, \alpha, \nu)$ are $(-5.70, 0.116, 0.085, 2.39, 0.40)$. There are
in fact two ST lines in the left plot, both with slope 0.116, but with slightly
different intercepts, depending on the adopted adjustment for the $\mathbb{E}\{\varepsilon\}$ term,
namely the estimated median of ε and the pseudo-mean $\tilde{\mu}$, as discussed
in § 4.3.4; the corresponding intercepts are -5.57 and -5.52, respectively.
Graphically, both ST lines are barely distinguishable from the MM line.

This example, similar to the earlier example of Figure 4.10, illustrates
how the wide possibility of regulating both skewness and tail thickness of
the ST distribution turns into the ability to down-weight points far out from
the bulk of the data, those termed 'outliers' in the terminology of robust
methods.

In this sense, the adoption of a highly flexible family of distributions to
describe data variability can be regarded as a viable approach to robustness.

The idea of employing a likelihood function which incorporates a form of robustness by allowing for adjustable tails has been present in the literature for a long time, as recalled in Section 4.1, but initial formulations have considered only symmetric densities. The possibility of employing distributions where one can regulate both skewness and kurtosis has been sketched briefly by Azzalini (1986), in connection with the introduction of asymmetric Subbotin distributions. An extensive exploration along these lines has been carried out by Azzalini and Genton (2008), taking examples from a range of areas: linear models, time series, multivariate analysis and classification. On the grounds of these numerical findings, as well as considerations similar to those discussed here in § 4.3.2, they have advocated the use of the ST distribution as a wide-purpose probability model, exhibiting good robustness properties in a range of diverse problems. Additional numerical illustration of the ST will be provided in the multivariate context of Chapter 6.

The inferential procedures so derived, whether Fisherian or Bayesian, do not of course descend from the principles of the canonical robustness approach, as formulated for instance in the classical works of Huber (1981; 2nd edition with E. M. Ronchetti, 2009) and Hampel *et al.* (1986). Hence, M-estimates, which constitute the most developed family of methods in that approach, must inevitably exhibit superior robustness properties, since they have been designed to be optimal from that viewpoint. However, there is no evidence in this direction emerging from the previous numerical examples.

To explore this question, consider the following simulation experiment. Sets of n data pairs are sampled from the simple regression scheme $y = \beta_0 + \beta_1 x + \varepsilon$, where the n components of the error term ε are generated from a mixture of $N(0,1)$ and $N(\Delta, 3^2)$ variates with weights $1 - \pi$ and π, respectively, under independence among the components of ε, and the vector x is made up of equally spaced values in $(0, 10)$. In the subsequent numerical work, we have chosen $n = 100$, $\beta_0 = 0$, $\beta_1 = 2$, π is either 0.05 or 0.10, and Δ ranges on 2.5, 5, 7.5 and 10. For each set of n data points, we compute these estimates of β_0 and β_1: least squares (LS), MM-estimates, least trimmed squares (LTS) robust regression, and MLE under assumption of ST distribution of ε. These steps have been replicated 50,000 times, followed by computation of the root mean square (RMS) estimation error of β_0 and β_1. The outcome is summarized in Figure 4.12, where the RMS error is plotted versus Δ for each combination of the β_j's and of the π values.

A first remark on Figure 4.12 is that the ST assumption for the error

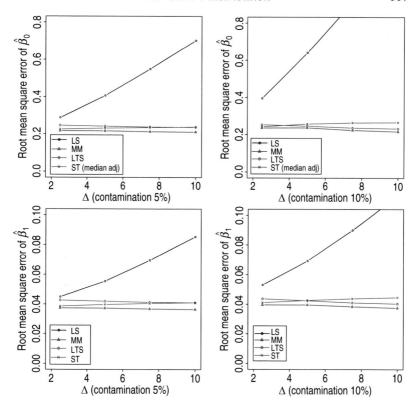

Figure 4.12 Root mean square error in the estimation of the coefficients $\beta_0 = 0$ (top plots) and $\beta_1 = 2$ (bottom plots) of a simple regression line when the standard normal error distribution has a 5% (left) or 10% (right) contamination from $N(\Delta, 3^2)$. Four estimation methods are evaluated; see the text for their description.

component produces sensible outcomes even if the ε's do not have ST distribution. Moreover, the estimates derived under this assumption compare favourably with the canonical robust methods, and they can perform even better than the LTS estimates if Δ is not large. As Δ increases, the relative performance of the estimates under ST assumption lowers slightly, especially in comparison to MM-estimates, but the difference is small. The visible pattern is that this loss will increase as Δ diverges, but only very gently, and we are already quite far from the main body of the data, namely 10 standard deviations of the main component of the error distribution. Recall that empirical studies mentioned in § 4.1 indicate that outlying observations in real data are (i) quite often asymmetrically placed, and (ii) seldom as extreme as those employed in many simulation studies.

As compensation for a limited loss in efficiency, ST can offer an advantage over M-estimates, thanks to the fact that we are working with a fully specified statistical model. This means that, once the ST parameters have been estimated, one can address in a simple way a question like 'what is $\mathbb{P}\{\varepsilon > c\}$, for some given c?'. This is not feasible when the estimates are defined directly by their estimating equation, as is the case for M-estimates.

An even more important question from a logical viewpoint is: 'what are we estimating?'. With the ST distribution, like for any other fully specified parametric formulation, it is clear that we estimate the parameters of the model. With M-estimation, the quantities to which the estimates converge asymptotically are given implicitly as the solution of some non-linear equation; see formula (B-3) and Theorem 6.4 on pp. 129–130 of Huber and Ronchetti (2009). In a simple location problem, if both the psi-function and the actual error distribution are symmetric about 0, then it is immediate to say that the solution of that equation is the location parameter of interest. In other cases, specifically when symmetry fails, no similar simple statement can be made.

Therefore, one can reasonably decide to accept a small loss on the side of efficiency to gain a plainer interpretability of the final outcome.

In contrast, it can be argued that inference based on a parametric model is valid insofar as the set of distributions comprising the model includes, ideally, the real distribution F^* which generates the data or, more plausibly, it 'nearly' does. It is known that asymptotically the MLE converges to the point θ_0 in the parameter space which corresponds to the minimum Kullback–Leibler divergence between F^* and the set of distributions in the parametric class. The adoption of a highly flexible family of distributions, such as the ST, can keep this divergence small in a vast set of situations, but of course it does not cover all possible cases, while the classical robust methods are designed to offer protection against 'all directions'. For instance, if the actual data distribution exhibits a pronounced bimodality, some route other than the ST must be followed, perhaps a mixture of two such components.

Preliminary inspection of the data and some understanding of the phenomenon under study remain essential components of the data analysis process, to avoid unreasonable usage of the available methods, no matter how powerful and 'optimal' they may be.

4.3.6 A real problem application

The aim of Walls (2005) is to predict the revenues of the film industry on the basis of some characteristics of individual movies, such as genre, rating category, year of production and so on. A prominent aspect of this context is that, in the author's words, 'the motion-picture market has a winner-take-all property where a small proportion of successful films earns the majority of box-office revenue'. This has motivated the adoption of very long-tailed distributions to handle data of this type, in particular the Pareto and the Lévy-stable distributions. While attractive because supported by theoretical justifications, these options are computationally demanding when embedded in a regression context, especially so the Lévy-stable distribution. This motivates consideration of alternatives, under the following requirements: 'To be useful in practice a statistical model of film returns should capture (1) the asymmetry implied by the winner-take-all property, (2) the heavy tails implied by the importance of extreme events, and (3) allow returns to be conditioned on a vector of explanatory variables'.

For a set of 1989 movies released on the North-American market from 1985 to 1996, Walls introduces a regression model for the log-transformed revenues as a linear function of the covariates and considers various alternative fitting procedures; these include standard least squares, minimum absolute deviation regression and maximum likelihood estimation assuming SN and ST distribution of the error term.

After a set of comparative remarks on the outcomes from the alternative methods, in the closing section the author writes: 'The skew-*t* regression model is particularly appealing in economics and finance where the data are characterized by heavy tails and skewness, and where interest is in analysing conditional distributions. However, the skew-*t* model is intuitively appealing in that it extends the Normal distribution by permitting tails that are heavy and asymmetric. Also, the skew-*t* model is computationally straightforward and estimable using standard statistical software that is freely available. In this respect, the skew-*t* model appears to be a practical approximation to the computationally overwhelming asymmetric Lévy-stable regression model.'

4.4 Complements

Complement 4.1 (Asymmetric Subbotin distribution of type I)   A quite natural choice for G_0 in (4.5) is the distribution function F_ν of the Subbotin density; see (4.3). Hence consider

$$f(x) = 2\frac{c_v}{\omega}\exp\left(-\frac{|z|^v}{v}\right)F_v(\alpha z), \qquad z = \frac{x - \xi}{\omega}, \qquad (4.25)$$

called the asymmetric Subbotin density of type I, briefly AS1. The graphical behaviour of this density exhibits little difference with respect to Figure 4.2; so the choice between the two alternatives has to be based on other considerations, such as mathematical convenience.

If Z denotes a random variable of type (4.25) having $\xi = 0$ and $\omega = 1$, it can be shown by direct integration that its mth moment is

$$\mathbb{E}\{Z^m\} = \frac{v^{m/v}\,\Gamma((m+1)/v)}{\Gamma(1/v)} \times \begin{cases} 1 & \text{if } m \text{ is even,} \\ \operatorname{sgn}(\alpha)\,I_\Delta(1/v, (m+1)/v) & \text{if } m \text{ is odd,} \end{cases}$$
$$(4.26)$$

where $\Delta = |\alpha|^v/(1 + |\alpha|^v)$ and $I_x(a, b)$ denotes the incomplete Beta function.

In case m is odd and $k = (m + 1)/v$ is an integer, the incomplete Beta function can be expressed as a finite sum, leading to

$$\mathbb{E}\{Z^m\} = \alpha\,\frac{2\,c_v\,v^{k-1}\,(k-1)!}{\Gamma(1/v)}\sum_{j=0}^{k-1}\frac{\Gamma(1/v + j)}{j!\,(1 + |\alpha|^v)^{j+1/v}}.$$

Complement 4.2 (Skew-Cauchy distribution)   An important special case of (4.11) occurs when $v = 1$, leading to a form of skew-Cauchy distribution. Since the distribution function of the Student's t on 2 d.f. is known to be

$$T(x; 2) = \frac{1}{2}\left(1 + \frac{x}{\sqrt{2 + x^2}}\right),$$

the skew-Cauchy density function is

$$t(x; \alpha, 1) = \frac{1}{\pi(1 + x^2)}\left(1 + \frac{\alpha x}{\sqrt{1 + (1 + \alpha^2)\,x^2}}\right), \qquad (4.27)$$

whose graphical appearance for two values of α is visible in the curves of Figure 4.4 with $v = 1$.

This distribution has been mentioned briefly by Gupta *et al.* (2002) and independently studied in detail by Behboodian *et al.* (2006), who have shown that the distribution function of (4.27) has the simple expression

$$T(x; \alpha, 1) = \frac{1}{\pi}\left(\arctan x + \arccos\frac{\delta(\alpha)}{\sqrt{1 + x^2}}\right), \qquad (4.28)$$

where $\delta(\alpha)$ is as in (2.6) on p. 26. Explicit inversion of this function is possible, providing the quantile function

$$T^{-1}(p; \alpha, 1) = \delta(\alpha) \sec\{\pi(p - \tfrac{1}{2})\} + \arctan\{\pi(p - \tfrac{1}{2})\},$$

for $p \in (0, 1)$. On setting $p = \tfrac{1}{2}$, we obtain that the median is $\delta(\alpha)$.

Other forms of skew-Cauchy distribution will be presented in the multivariate context; see Complement 6.3.

Complement 4.3 (ST distribution function) Jamalizadeh *et al.* (2009a) have shown that the distribution function $T(x; \alpha, \nu)$ of the skew-*t* distribution (4.11) satisfies the recursive relationship

$$T(x; \alpha, \nu + 1) = \frac{\Gamma\left(\tfrac{1}{2}\nu\right) (\nu + 1)^{(\nu-1)/2}}{\sqrt{\pi}\,\Gamma\left(\tfrac{1}{2}(\nu + 1)\right)} \frac{x}{(\nu + 1 + x^2)^{\nu/2}} T\left(\frac{\sqrt{\nu}\,\alpha\,x}{\sqrt{\nu + 1 + x^2}}; \nu\right)$$

$$+ T\left(\sqrt{\frac{\nu - 1}{\nu + 1}}\, x; \alpha, \nu - 1\right) \tag{4.29}$$

for $\nu > 1$, possibly non-integer. Combination of (4.29), (4.28) and the distribution function for $\nu = 2$, which is

$$T(x; \alpha, 2) = \frac{1}{2} - \frac{1}{\pi} \arctan \alpha + \frac{x}{\sqrt{2 + x^2}} \left(\frac{1}{2} + \frac{1}{\pi} \arctan \frac{\alpha x}{\sqrt{2 + x^2}}\right),$$

allows us to compute the skew-*t* distribution function for $\nu = 3, 4, \ldots$ On setting $\alpha = 0$, (4.29) lends a simplified recursion for the regular *t* distribution function $T(x; \nu)$.

Complement 4.4 (ST tail behaviour) To study the tail behaviour of the ST distribution function, start by rewriting the density (4.11) as

$$t(x; \alpha, \nu) = 2 \frac{\Gamma\left(\tfrac{1}{2}(\nu + 1)\right)}{\Gamma\left(\tfrac{1}{2}\nu\right)\sqrt{\pi\nu}} x^{-(\nu+1)} \left(\frac{1}{x^2} + \frac{1}{\nu}\right)^{-\frac{\nu+1}{2}} T(w(x); \nu + 1),$$

where $w(x)$, given by (4.12), converges to $\mathrm{sgn}(x)\,\alpha\,\sqrt{\nu + 1}$ when $|x|$ diverges. It then follows that

$$t(x; \alpha, \nu) \sim c_{\alpha,\nu}\, |x|^{-(\nu+1)} \qquad \text{as} \qquad x \to \infty, \tag{4.30}$$

where the symbol '$\sim$' denotes asymptotic equivalence, that is, the ratio of the two sides converges to unity as x diverges, and

$$c_{\alpha,\nu} = 2 \frac{\Gamma\left(\tfrac{1}{2}(\nu + 1)\right) \nu^{\nu/2}}{\Gamma\left(\tfrac{1}{2}\nu\right)\sqrt{\pi}} T(\mathrm{sgn}(x)\,\alpha\,\sqrt{\nu + 1}; \nu + 1).$$

An implication of (4.30) is that the ST density decays at the same rate as the regular Student's t, whose limiting behaviour corresponds to the value $c_{0,v}$ in (4.30). By integrating (4.30), we obtain the following approximations for the right and left tail probabilities:

$$1 - T(x; \alpha, v) \sim \frac{c_{\alpha,v}}{v} x^{-v} \quad \text{as} \quad x \to +\infty,$$

$$T(x; \alpha, v) \sim \frac{c_{\alpha,v}}{v} |x|^{-v} \quad \text{as} \quad x \to -\infty,$$

if $\alpha \geq 0$. If $\alpha < 0$, recall that $T(x; -\alpha, v) = T(-x; \alpha, v)$. A more formal argument leading to an asymptotically equivalent expression is given by Padoan (2011, p. 980), once a typographical error is corrected in the quoted degrees of freedom of the function T in the expression corresponding to $c_{\alpha,v}$ here.

Using Theorem 1.6.2 in Leadbetter *et al.* (1983), one can establish that the domain of attraction of $T(x; \alpha, v)$ is, like in the symmetric case, the Fréchet family of distributions, since for any $a > 0$,

$$\lim_{x \to \infty} \frac{1 - T(a\,x; \alpha, v)}{1 - T(x; \alpha, v)} = a^{-v}.$$

For more information on these aspects, see Chang and Genton (2007) and Padoan (2011).

Complement 4.5 (Tests for normality within the ST class)   To test that a sample $y_1, \ldots, y_n$ drawn from a ST variable is actually of Gaussian type, Carota (2010) introduces the transformed parameter $\theta = (\alpha/v, 1/v)$ and considers the score test for the null hypothesis that $\theta = (0, 0)$. The adoption of this parameterization is a technical device adopted by the author to facilitate the computation of the score test as the point of interest. Since the value $1/v = 0$ lies on the boundary of the parameter space, the test statistic must be considered for a value $1/v = \varepsilon > 0$, followed by a limit operation as $\varepsilon \to 0$. The author obtains that the score test statistic for normality within the ST class is

$$S \approx \frac{n}{6} \bar{\gamma}_1^2 + \frac{n}{24} \bar{\gamma}_2^2,$$

where $\bar{\gamma}_1$ is the sample coefficient of skewness (3.11) and

$$\bar{\gamma}_2 = \frac{\sum_i (y_i - \bar{y})^4 / n}{s^4} - 3 \tag{4.31}$$

is the sample coefficient of excess kurtosis; s is as defined in (3.5). The above approximation to S is a familiar quantity, commonly referred to in the econometric literature as the Jarque–Bera test statistic.

Problems

4.1 Churchill (1946) examines the following density function, attributing its discovery to Stieltjes:

$$f(x) = \frac{1}{48}\left(1 - \operatorname{sgn}(x)\sin|x|^{1/4}\right)\exp\left(-|x|^{1/4}\right), \qquad -\infty < x < \infty.$$

Show that its pth moment is $\Gamma\{4(p+1)\}/6$ if p is even, and 0 if p is odd. For the latter statement, it helps to take into account that

$$\int_0^\infty x^{q-1} e^{-x} \sin x \, dx = 2^{-q/2} \Gamma(q) \sin(q\pi/4).$$

Since all odd moments are zero, so is the coefficient of skewness γ_1, in spite of the visually striking asymmetry of $f(x)$ – plot it to convince yourself! Show that $f(x)$ is of type (1.3) with base density f_0 of Subbotin type (4.1), up to an inessential reparameterization.

4.2 Prove (4.26).

4.3 Setting $v = 1$ in (4.25) lends a form of asymmetric Laplace distribution. Show that its moments are

$$\mathbb{E}\{Z^m\} = m! \times \begin{cases} 1 & \text{if } m \text{ is even,} \\ \operatorname{sgn}(\alpha)\left(1 - \dfrac{1}{(1+|\alpha|)^{m+1}}\right) & \text{if } m \text{ is odd,} \end{cases}$$

(Azzalini, 1986).

4.4 Prove that the density (4.25) is log-concave if $v \geq 1$ (Azzalini, 1986).

4.5 Check expressions (4.16)–(4.19).

4.6 Show that the limit behaviour of b_v defined by (4.15) is

$$b_v \sim \sqrt{\frac{2}{\pi}}\left(1 + \frac{3}{4v} + \frac{25}{32v^2}\right) \qquad \text{as } v \to \infty.$$

4.7 Prove the statement in the text that ST density (4.11) is unimodal (Capitanio, 2012, once the result is restricted to the univariate case).

5

The multivariate skew-normal distribution

5.1 Introduction

5.1.1 Definition and basic properties

A quite natural and simple extension of the skew-normal density (2.1) to the d-dimensional case, still of type (1.2), is given by

$$\varphi_d(x; \bar{\Omega}, \alpha) = 2\,\varphi_d(x; \bar{\Omega})\,\Phi(\alpha^\top x), \qquad x \in \mathbb{R}^d, \qquad (5.1)$$

where $\bar{\Omega}$ is a positive-definite $d \times d$ correlation matrix, $\varphi_d(x; \Sigma)$ denotes the density function of a $N_d(0, \Sigma)$ variate and α is the d-dimensional vector parameter.

There are many other types of multivariate skew-normal distribution we might consider, some of which will indeed be examined later in this book. As already said, (5.1) represents what arguably is the simplest option involving a modulation factor of Gaussian type operating on a multivariate normal base density.

We shall refer to a variable Z with density (5.1) as a 'normalized' *multivariate skew-normal* variate. For applied work, we need to introduce location and scale parameters via the transformation

$$Y = \xi + \omega Z, \qquad (5.2)$$

where $\xi \in \mathbb{R}^d$ and $\omega = \mathrm{diag}(\omega_1, \ldots, \omega_d) > 0$, leading to the general form of multivariate SN variables. It is immediate that the density function of Y at $x \in \mathbb{R}^d$ is

$$\det(\omega)^{-1}\varphi_d(z; \bar{\Omega}, \alpha) = 2\,\varphi_d(x - \xi; \Omega)\,\Phi(\alpha^\top \omega^{-1}(x - \xi)), \qquad (5.3)$$

where $z = \omega^{-1}(x - \xi)$ and $\Omega = \omega\bar{\Omega}\omega$. We write $Y \sim \mathrm{SN}_d(\xi, \Omega, \alpha)$ and the parameter components will be called location, scale matrix and slant, respectively. When this notation is used, we shall be implicitly assuming that $\Omega > 0$. Note that ω can be written as

$$\omega = (\Omega \odot I_d)^{1/2}$$

124

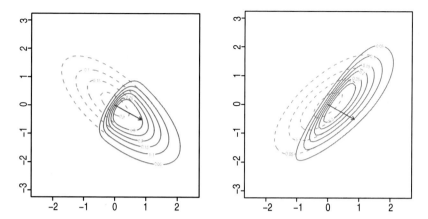

Figure 5.1 Contour plot of two bivariate skew-normal density functions when $\xi = (0,0)$, $\alpha = (5,-3)^\top$, $\Omega_{11} = 1$, $\Omega_{22} = 1$ and $\Omega_{12} = -0.7$ in the left-side panel, and $\Omega_{12} = 0.7$ in the right-side panel. In each panel the dashed grey line represents the contour plot of the corresponding modulated bivariate normal distribution, and the arrow represents the vector α divided by its Euclidean norm.

where $\odot$ denotes the entry-wise or Hadamard product. We shall use this type of notation repeatedly in the following.

The shape of the multivariate SN density depends on the combined effect of Ω and α. For the bivariate case, a graphical illustration of the interplay between these components is provided in Figure 5.1, which shows the contour plots of two densities having the same parameter set except Ω_{12}. Here the corresponding base densities f_0, displayed by dashed grey lines, are formed by the reflection of each other with respect to the vertical axis, but the modulation effect produced by the same α leads to quite different densities.

Many properties of the univariate skew-normal distribution extend directly to the multivariate case. These are the simplest ones:

$$\varphi_d(x; \Omega, 0) = \varphi_d(x; \Omega), \quad \text{for all } x, \tag{5.4}$$

$$\varphi_d(0; \Omega, \alpha) = \varphi_d(0; \Omega), \tag{5.5}$$

$$-Z \sim \mathrm{SN}_d(0, \bar{\Omega}, -\alpha), \tag{5.6}$$

$$(Y - \xi)^\top \Omega^{-1}(Y - \xi) = Z^\top \bar{\Omega}^{-1} Z \sim \chi_d^2, \quad \text{for all } \alpha, \tag{5.7}$$

where Z has distribution (5.1) and Y is given by (5.2).

Proposition 5.1 *The* $\mathrm{SN}_d(\xi, \Omega, \alpha)$ *density (5.3) is log-concave, i.e., its logarithm is a concave function of x, for any choice of the parameters.*

The proof is a simple extension of Proposition 2.6 for the univariate case; see Problem 5.1. From this we conclude that the regions delimited by contour lines of the density are convex sets, and, of course, the mode is unique.

Before entering more technical aspects, some remarks on the choice of parameterization are appropriate. For algebraic simplicity, one might think of replacing $\omega^{-1}\alpha$ in the final factor of (5.3) by a single term η, say, and view the distribution as a function of (ξ, Ω, η). While use of η does simplify several expressions, and we shall make use of it at places, its adoption for parameterizing the family is questionable, since η reflects both the shape and the scale of the distribution. This choice would be similar to expressing the linear dependence between two variables via their covariance instead of their correlation.

Another notation in use replaces ω^{-1} in (5.3) by $\Omega^{-1/2}$. In this case the problem is that there are many possible options for the square root of Ω, leading to actually different densities, and there is no decisive reason for choosing one specific alternative.

5.1.2 Moment generating function

The following lemma is an immediate extension of Lemma 2.2; the proof follows by simply noticing that $h^\top U \sim \mathrm{N}(0, h^\top \Sigma h)$ if $U \sim \mathrm{N}_d(0, \Sigma)$. The subsequent statement illustrates the technique of 'completing the square' for a skew-normal type of integrand.

Lemma 5.2 *If* $U \sim \mathrm{N}_d(0, \Sigma)$ *then*

$$\mathbb{E}\{\Phi(h^\top U + k)\} = \Phi\left(\frac{k}{\sqrt{1 + h^\top \Sigma h}}\right), \qquad h \in \mathbb{R}^d, \, k \in \mathbb{R}. \qquad (5.8)$$

Lemma 5.3 *If A is a symmetric positive definite* $d \times d$ *matrix, a and c are d-vectors and* c_0 *is a scalar, then*

$$I = \int_{\mathbb{R}^d} \frac{1}{(2\pi)^{d/2} \det(A)^{1/2}} \exp\left\{-\tfrac{1}{2}(x^\top A^{-1} x - 2a^\top x)\right\} \Phi(c_0 + c^\top x) \, dx$$

$$= \exp\left(\tfrac{1}{2}a^\top A a\right) \Phi\left(\frac{c_0 + c^\top A a}{\sqrt{1 + c^\top A c}}\right). \qquad (5.9)$$

Proof In the integrand of I rewrite $x^\top A^{-1} x - 2a^\top x$ as $(x-\mu)^\top A^{-1}(x-\mu) - \mu^\top A^{-1}\mu$ where $\mu = Aa$, so that

$$I = \exp(\tfrac{1}{2}a^\top A a) \int_{\mathbb{R}^d} \varphi(y; A) \, \Phi\{c_0 + c^\top(y + \mu)\} \, dy$$

after a change of variable. Use of Lemma 5.2 gives (5.9). QED

To compute the moment generating function $M(t)$ of $Y \sim SN_d(\xi, \Omega, \alpha)$, write $Y = \xi + \omega Z$, where $Z \sim SN_d(0, \bar{\Omega}, \alpha)$. Then, using Lemma 5.3, we obtain

$$M(t) = \exp(t^\top \xi) \int_{\mathbb{R}^d} 2 \exp(t^\top \omega z)\, \varphi_d(z; \bar{\Omega})\, \Phi(\alpha^\top z)\, dz$$

$$= 2 \exp(t^\top \xi + \tfrac{1}{2} t^\top \Omega t)\, \Phi(\delta^\top \omega t), \qquad t \in \mathbb{R}^d, \qquad (5.10)$$

where

$$\delta = \left(1 + \alpha^\top \bar{\Omega} \alpha\right)^{-1/2} \bar{\Omega} \alpha. \qquad (5.11)$$

For later use, we write down the inverse relationship:

$$\alpha = \left(1 - \delta^\top \bar{\Omega}^{-1} \delta\right)^{-1/2} \bar{\Omega}^{-1} \delta. \qquad (5.12)$$

A simple corollary obtained using the above expression of $M(t)$ is the next statement, which is the multivariate extension of Proposition 2.3.

Proposition 5.4 *If $Y_1 \sim SN_d(\xi, \Omega, \alpha)$ and $Y_2 \sim N_d(\mu, \Sigma)$ are independent variables, then*

$$X = Y_1 + Y_2 \sim SN_d(\xi + \mu, \Omega_X, \tilde{\alpha}),$$

where

$$\Omega_X = \Omega + \Sigma, \qquad \tilde{\alpha} = \left(1 + \eta^\top \Omega_X^{-1} \eta\right)^{-1/2} \omega_X \Omega_X^{-1} \Omega \eta,$$

having set $\eta = \omega^{-1} \alpha$ and $\omega_X = (\Omega_X \odot I_d)^{1/2}$.

Similarly to the univariate case, it can be shown that, when both summands Y_1 and Y_2 are 'proper' independent SN variates, that is with non-null slant, their sum is not SN. The proof is, in essence, the same as Proposition 5.5, given later. The only formal difference is in the leading sign of quadratic forms appearing in the $\exp(\cdot)$ terms, but this does not affect the argument.

5.1.3 Stochastic representations

Conditioning and selective sampling

Specification of (1.9)–(1.11) to the present context says that, if $X_0 \sim N_d(0, \bar{\Omega})$ and $T \sim N(0, 1)$ are independent variables, then both

$$Z' = (X_0 | T > \alpha^\top X_0), \qquad Z = \begin{cases} X_0 & \text{if } T > \alpha^\top X_0, \\ -X_0 & \text{otherwise} \end{cases} \qquad (5.13)$$

have distribution $SN_d(0, \bar{\Omega}, \alpha)$.

This scheme can be rephrased in an equivalent form which allows an interesting interpretation. Define

$$X_1 = \left(1 + \alpha^\top \bar{\Omega} \alpha\right)^{-1/2} \left(\alpha^\top X_0 - T\right)$$

such that

$$X = \begin{pmatrix} X_0 \\ X_1 \end{pmatrix} \sim N_{d+1}(0, \Omega^*), \qquad \Omega^* = \begin{pmatrix} \bar{\Omega} & \delta \\ \delta^\top & 1 \end{pmatrix}, \tag{5.14}$$

where δ is given by (5.11) and Ω^* is a full-rank correlation matrix. Then the variables in (5.13) can be written as

$$Z' = (X_0 | X_1 > 0), \qquad Z = \begin{cases} X_0 & \text{if } X_1 > 0, \\ -X_0 & \text{otherwise.} \end{cases} \tag{5.15}$$

For random number generation the second form is preferable, as discussed in Complement 1.1. However, the first form of (5.15) is interesting for its qualitative interpretation, since it indicates a link between the skew-normal distribution and a censoring mechanism, fairly common in an applied context, especially in the social sciences, where a variable X_0 is observed only when a correlated variable, X_1, which is usually unobserved, fulfils a certain condition. This situation is commonly referred to as selective sampling.

Another use of the first form of (5.15) allows us to express the distribution function of a multivariate SN variable, but we defer this point to the slightly more general case of § 5.3.3.

Additive representation

Consider a $(d+1)$-dimensional normal variable U which is partitioned into components U_0 and U_1 of dimensions d and 1, respectively, such that the joint distribution is

$$U = \begin{pmatrix} U_0 \\ U_1 \end{pmatrix} \sim N_{d+1}\left(0, \begin{pmatrix} \bar{\Psi} & 0 \\ 0 & 1 \end{pmatrix}\right), \tag{5.16}$$

where $\bar{\Psi}$ is a full-rank correlation matrix. Given a vector $\delta = (\delta_1, \dots, \delta_d)^\top$ with all elements in $(-1, 1)$, define similarly to (2.14)

$$Z_j = \left(1 - \delta_j^2\right)^{1/2} U_{0j} + \delta_j |U_1|, \tag{5.17}$$

for $j = 1, \dots, d$. If $Z_1, \dots, Z_d$ are arranged in a d-vector Z and we set

$$D_\delta = \left(I_d - \text{diag}(\delta)^2\right)^{1/2}, \tag{5.18}$$

we can write more compactly

$$Z = D_\delta \, U_0 + \delta \, |U_1| . \tag{5.19}$$

Some algebraic work says that Z has a d-dimensional skew-normal distribution with parameters $(\bar{\Omega}, \alpha)$ related to δ and $\bar{\Psi}$ as follows:

$$\lambda = D_\delta^{-1} \delta , \tag{5.20}$$

$$\bar{\Omega} = D_\delta \, (\bar{\Psi} + \lambda \lambda^\top) \, D_\delta , \tag{5.21}$$

$$\alpha = \left(1 + \lambda^\top \bar{\Psi}^{-1} \lambda \right)^{-1/2} D_\delta^{-1} \bar{\Psi}^{-1} \lambda ; \tag{5.22}$$

see Problem 5.2. In the scalar case, $\bar{\Omega}$ and $\bar{\Psi}$ reduce to 1 and $\lambda = \alpha$.

A direct link between the ingredients of the additive representation and those in (5.14) can be established by the standard orthogonalization scheme

$$U_1 = X_1, \quad U_0' = X_0 - \mathbb{E}\{X_0|X_1\} = X_0 - \delta \, X_1 \sim \mathrm{N}_d(0, \; \bar{\Omega} - \delta \, \delta^\top) \tag{5.23}$$

which, after transformation $U_0 = D_\delta^{-1} U_0'$ to have unit variances, leads to (5.16). Inversion of these relationships shows how to obtain X from U.

Minima and maxima

To introduce a stochastic representation in the form of minima and maxima, which generalizes the analogous one for the scalar case presented in Chapter 2, we make use of the variables and other elements introduced in the previous paragraph. Note that Z_j in (5.17) is algebraically equivalent to

$$Z_j = \mathrm{sgn}(\delta_j) \tfrac{1}{2} \, |V_j - W_j| + \tfrac{1}{2} (V_j + W_j), \qquad j = 1, \ldots, d,$$

where

$$V_j = (1 - \delta_j^2)^{1/2} \, U_{0j} + \delta_j \, U_1, \qquad W_j = (1 - \delta_j^2)^{1/2} \, U_{0j} - \delta_j \, U_1 .$$

The joint distribution of $V = (V_1, \ldots, V_d)$ and $W = (W_1, \ldots, W_d)$ is singular Gaussian, specifically

$$\begin{pmatrix} V \\ W \end{pmatrix} \sim \mathrm{N}_{2d}\left(0, \begin{pmatrix} D_\delta \bar{\Psi} D_\delta + \delta \delta^\top & D_\delta \bar{\Psi} D_\delta - \delta \delta^\top \\ D_\delta \bar{\Psi} D_\delta - \delta \delta^\top & D_\delta \bar{\Psi} D_\delta + \delta \delta^\top \end{pmatrix} \right), \tag{5.24}$$

where $\delta = (\delta_1, \ldots, \delta_d)^\top$ and D_δ is as in (5.18). The variables

$$V - W \sim \mathrm{N}_d(0, 4 \delta \delta^\top), \qquad V + W \sim \mathrm{N}_d(0, 4 D_\delta \bar{\Psi} D_\delta)$$

are independent, and the equality $V - W = 2U_1 \delta$ confirms that $V - W$ has singular distribution. Recalling that

$$\max\{a, b\} = \tfrac{1}{2} |a - b| + \tfrac{1}{2} (a + b), \quad \min\{a, b\} = -\tfrac{1}{2} |a - b| + \tfrac{1}{2} (a + b)$$

and writing

$$
Z_j = \begin{cases} \max\{V_j, W_j\} & \text{if } \delta_j \geq 0, \\ \min\{V_j, W_j\} & \text{otherwise,} \end{cases} \tag{5.25}
$$

it is visible that $Z = (Z_1, \ldots, Z_d)^\top \sim SN_d$ with parameters (5.21)–(5.22).

5.1.4 Marginal distributions and another parameterization

Closure of the SN family with respect to marginalization follows from (5.10). More specifically, suppose that $Y \sim SN_d(\xi, \Omega, \alpha)$ is partitioned as $Y^\top = (Y_1^\top, Y_2^\top)$ where the two components have dimension h and $d - h$, respectively, and denote by

$$
\xi = \begin{pmatrix} \xi_1 \\ \xi_2 \end{pmatrix}, \quad \Omega = \begin{pmatrix} \Omega_{11} & \Omega_{12} \\ \Omega_{21} & \Omega_{22} \end{pmatrix}, \quad \alpha = \begin{pmatrix} \alpha_1 \\ \alpha_2 \end{pmatrix}, \quad \delta = \begin{pmatrix} \delta_1 \\ \delta_2 \end{pmatrix} \tag{5.26}
$$

the corresponding partitions of ξ, Ω, α and δ. Evaluation of (5.10) at $t = (s^\top, 0)^\top$ gives the moment generating function of Y_1 as

$$
M_{Y_1}(s) = 2 \exp\left(s^\top \xi_1 + \tfrac{1}{2} s^\top \Omega_{11} s\right) \Phi(\delta_1^\top \omega_{11} s), \qquad s \in \mathbb{R}^h,
$$

showing that Y_1 is of skew-normal type with location ξ_1 and scale matrix Ω_{11}. To find the slant parameter, $\alpha_{1(2)}$ say, we use (5.12) with δ replaced by δ_1, the first h components of (5.11). After some algebra, we arrive at

$$
\alpha_{1(2)} = \left(1 + \alpha_2^\top \bar{\Omega}_{22 \cdot 1} \alpha_2\right)^{-1/2} \left(\alpha_1 + \bar{\Omega}_{11}^{-1} \bar{\Omega}_{12} \alpha_2\right) \tag{5.27}
$$

where

$$
\bar{\Omega}_{11}^{-1} = (\bar{\Omega}_{11})^{-1}, \qquad \bar{\Omega}_{22 \cdot 1} = \bar{\Omega}_{22} - \bar{\Omega}_{21} \bar{\Omega}_{11}^{-1} \bar{\Omega}_{12} \tag{5.28}
$$

on partitioning $\bar{\Omega}$ similarly to Ω. To conclude, marginally

$$
Y_1 \sim SN_h(\xi_1, \Omega_{11}, \alpha_{1(2)}). \tag{5.29}
$$

Some remarks on the interpretation of the parameters $(\bar{\Omega}, \alpha)$ are now appropriate. From (5.17) it is apparent that the entries of the vector δ of the joint distribution coincide with the δ parameters of the marginal distributions. The same fact is visible also from the above expression of $M_{Y_1}(t)$, on taking $h = 1$. On the contrary, the jth entry of α does not individually provide information on the jth marginal of the joint distribution. In fact, from α_j we cannot even infer the sign of the corresponding component δ_j, that is, whether the jth marginal is positively or negatively asymmetric. However, a meaning can be attached to a null value of α_j, as we shall see in § 5.3.2 and § 5.3.5.

If one wants a parameterization where the parameter components have an interpretation as an individual slant parameter, this is possible on the basis of $(\bar{\Psi}, \lambda)$, recalling (5.20). We have seen that each choice of $\bar{\Psi}$ and λ in (5.16)–(5.19) corresponds to a distribution of type (5.1). The converse also holds: for each choice of $(\bar{\Omega}, \alpha)$ there is a corresponding choice of $(\bar{\Psi}, \delta)$ or equivalently of $(\bar{\Psi}, \lambda)$, in (5.16)–(5.19), leading to the same distribution; see Problem 5.3. Hence $(\bar{\Omega}, \alpha)$ and $(\bar{\Psi}, \lambda)$ are equivalent parameterizations for the same set of distributions. In both cases, the two components are variation independent, that is, they can be selected independently of each other. As an example of the contrary, $\bar{\Omega}$ and δ are not variation independent.

For the full class (5.3), write

$$\Omega = \psi(\bar{\Psi} + \lambda\lambda^{\top})\psi = \Psi + \psi\lambda\lambda^{\top}\psi$$

where $\psi = \omega D_\delta = D_\delta\omega$ now represents the scale factor and $\Psi = \psi\bar{\Psi}\psi$; here D_δ is as in (5.18). Hence (5.3) can be equivalently expressed via the (ξ, Ψ, λ) parameter set as

$$2\,\varphi_d(x - \xi; \Psi + \psi\lambda\lambda^{\top}\psi)\,\Phi\left(\frac{1}{\sqrt{1 + \lambda^{\top}\bar{\Psi}^{-1}\lambda}}\,\lambda^{\top}\bar{\Psi}^{-1}\psi^{-1}(x - \xi)\right). \qquad (5.30)$$

As already stated, the parameterization (ξ, Ψ, λ) has the advantage that the components of λ are interpretable individually. The reason why the parameterization (ξ, Ω, α) has been given a primary role is that it allows a simpler treatment in other respects. A basic fact is that (5.3) constitutes a simpler expression than (5.30). However, the reasons in favour of (Ω, α) are not indisputable, and one may legitimately prefer to use (Ψ, λ).

The skew-normal family is not closed under conditioning. A slight extension of the SN family which enjoys this property will be discussed in §5.3.

5.1.5 Cumulants and related quantities

From (5.10), the cumulant generating function of $Y \sim SN_d(\xi, \Omega, \alpha)$ is

$$K(t) = \log M(t) = \xi^{\top}t + \tfrac{1}{2}t^{\top}\Omega t + \zeta_0(\delta^{\top}\omega t), \qquad t \in \mathbb{R}^d,$$

where $\zeta_0(x)$ is defined by (2.18). Taking into account (2.19), the first two derivatives of $K(t)$ are

$$\frac{\mathrm{d}}{\mathrm{d}t}K(t) = \xi + \Omega t + \zeta_1(\delta^{\top}\omega t)\,\omega\,\delta,$$

$$\frac{\mathrm{d}^2}{\mathrm{d}t\,\mathrm{d}t^{\top}}K(t) = \Omega + \zeta_2(\delta^{\top}\omega t)\,\omega\,\delta\,\delta^{\top}\omega,$$

and their values at $t = 0$ give

$$\mu = \mathbb{E}\{Y\} = \xi + \omega b \delta = \xi + \omega \mu_z, \tag{5.31}$$

$$\Sigma = \mathrm{var}\{Y\} = \Omega - \omega \mu_z \mu_z^\top \omega = \omega \Sigma_z \omega \tag{5.32}$$

where, analogously to the univariate case in § 2.1.2, we have set

$$\mu_z = b \delta = \mathbb{E}\{Z\}, \qquad \Sigma_z = \bar{\Omega} - \mu_z \mu_z^\top = \mathrm{var}\{Z\}$$

for $Z \sim \mathrm{SN}_d(0, \bar{\Omega}, \alpha)$. If $\xi = 0$, a quick way to obtain that $\mathbb{E}\{Y Y^\top\} = \Omega$ is by simply recalling the modulation invariance property.

The rth-order derivative of $K(t)$, for $r > 2$, takes the form

$$\frac{\mathrm{d}^r}{\mathrm{d}t_i \, \mathrm{d}t_j \, \cdots \, \mathrm{d}t_h} K(t) = \zeta_r(\delta^\top \omega \, t) \, \omega_i \, \omega_j \cdots \omega_h \, \delta_i \, \delta_j \cdots \delta_h, \tag{5.33}$$

where the expression of $\zeta_r(x)$ up to $r = 4$ is given by (2.20).

Evaluation at $t = 0$ of the above derivatives allows us to obtain an explicit expression of the coefficients of multivariate skewness and kurtosis introduced by Mardia (1970, 1974). Specifically, evaluation of (5.33) at $t = 0$ for $r = 3$ and insertion in (1.1) of Mardia (1974) lead to

$$\gamma_{1,d}^M = \beta_{1,d}^M = \zeta_3(0)^2 \sum_{vst} \sum_{v's't'} \delta_v \delta_s \delta_t \delta_{v'} \delta_{s'} \delta_{t'} \sigma_z^{vv'} \sigma_z^{ss'} \sigma_z^{tt'}$$

$$= \left(\frac{4 - \pi}{2}\right)^2 \left(\mu_z^\top \Sigma_z^{-1} \mu_z\right)^3 \tag{5.34}$$

where $\Sigma_z^{-1} = (\sigma_z^{st})$, and similarly when $r = 4$ we obtain

$$\gamma_{2,d}^M = \beta_{2,d}^M - d(d + 2) = \zeta_4(0) \sum_{rstu} \delta_v \delta_s \delta_t \delta_u \sigma_z^{vs} \sigma_z^{tu}$$

$$= 2(\pi - 3) \left(\mu_z^\top \Sigma_z^{-1} \mu_z\right)^2 \tag{5.35}$$

from expressions (1.2) and (2.9) of Mardia (1974). The two measures, $\gamma_{1,d}^M$ and $\gamma_{2,d}^M$, depend on α and $\bar{\Omega}$ through the quadratic form $\mu_z^\top \Sigma_z^{-1} \mu_z$, which in turn can be rewritten as

$$\mu_z^\top \Sigma_z^{-1} \mu_z = \frac{(2/\pi) \alpha_*^2}{1 + (1 - 2/\pi) \alpha_*^2}, \tag{5.36}$$

where

$$\alpha_* = (\alpha^\top \bar{\Omega} \alpha)^{1/2} \in [0, \infty) \tag{5.37}$$

can then be seen as the regulating quantity. Therefore, as for Mardia's measures, the scalar quantity α_* encapsulates comprehensively the departure from normality. Equivalently, (5.36) and other expressions which will appear later can be written as functions of

$$\delta_* = (\delta^\top \bar{\Omega}^{-1} \delta)^{1/2} \in [0, 1), \tag{5.38}$$

where as usual δ is given by (5.11). These quantities are connected via

$$\delta_*^2 = \frac{\alpha_*^2}{1 + \alpha_*^2}, \qquad \alpha_*^2 = \frac{\delta_*^2}{1 - \delta_*^2}.$$

Some algebraic manipulation gives further insight about α_*. In (5.36) write α_* as a function $\delta(\alpha_*)$ according to (2.6) on p. 26. We can then rewrite (5.36) as $\mu_{\alpha_*}^2 / \sigma_{\alpha_*}^2$, where the two components are functions of $\delta(\alpha_*)$ given by (2.26). Finally, we arrive at

$$\gamma_{1,d}^M = \left(\frac{4 - \pi}{2}\right)^2 \left(\frac{\mu_{\alpha_*}^2}{\sigma_{\alpha_*}^2}\right)^3, \qquad \gamma_{2,d}^M = 2(\pi - 3)\left(\frac{\mu_{\alpha_*}^2}{\sigma_{\alpha_*}^2}\right)^2, \tag{5.39}$$

that is, $\gamma_{1,d}^M$ and $\gamma_{2,d}^M$ correspond to the square of γ_1 and to the γ_2 coefficient, respectively, for the distribution $SN(0, 1, \alpha_*)$. These expressions arise from mere algebraic rewriting of (5.34) and (5.35), but they are notionally associated with a distribution $SN(0, 1, \alpha_*)$. This idea will take a more precise shape in § 5.1.8.

An implication of (5.39) is that $\gamma_{1,d}^M$ and $\gamma_{2,d}^M$ range from 0 to $(\gamma_1^{max})^2$ and to γ_2^{max}, respectively, where γ_1^{max} and γ_2^{max} are given by (2.31).

5.1.6 Linear, affine and quadratic forms

From the moment generating function (5.10), it is visible that the family of multivariate skew-normal distributions is closed under affine transformations. More specifically, if $Y \sim SN_d(\xi, \Omega, \alpha)$, A is a full-rank $d \times h$ matrix, with $h \leq d$, and $c \in \mathbb{R}^h$, then some algebraic work shows that

$$X = c + A^\top Y \sim SN_h(\xi_X, \Omega_X, \alpha_X) \tag{5.40}$$

where

$$\xi_X = c + A^\top Y, \tag{5.41}$$
$$\Omega_X = A^\top \Omega A, \tag{5.42}$$
$$\alpha_X = \left(1 - \delta^\top \omega A \Omega_X^{-1} A^\top \omega \delta\right)^{-1/2} \omega_X \Omega_X^{-1} A^\top \omega \delta \tag{5.43}$$

having set $\omega_X = (\Omega_X \odot I_h)^{1/2}$ and, as usual, δ is given by (5.11). When $h = 1$, so that A reduces to a vector, a say, (5.43) simplifies to

$$\alpha_X = \left(a^\top \Omega a - (a^\top \omega \delta)^2\right)^{-1/2} a^\top \omega \delta. \tag{5.44}$$

To examine the question of independence among components of an SN variable, we need the following preliminary result, which is also of independent interest.

Proposition 5.5 *For any choice of $a_1, a_2 \in \mathbb{R}$, $\mu_1, b_1 \in \mathbb{R}^p$, $\mu_2, b_2 \in \mathbb{R}^q$ such that $b_1 \neq 0$ and $b_2 \neq 0$ and symmetric positive-definite matrices Σ_1, Σ_2, there exist no $a, c \in \mathbb{R}$, $b, \mu \in \mathbb{R}^{p+q}$ and matrix Σ such that*

$$\varphi_p(x_1 - \mu_1; \Sigma_1)\, \Phi(a_1 + b_1^\top x_1)\, \varphi_q(x_2 - \mu_2; \Sigma_2)\, \Phi(a_2 + b_2^\top x_2)$$
$$= c\, \varphi_{p+q}(x - \mu; \Sigma)\, \Phi(a + b^\top x) \tag{5.45}$$

for all $x_1 \in \mathbb{R}^p$, $x_2 \in \mathbb{R}^q$, $x = (x_1^\top, x_2^\top)^\top$.

Proof Select one non-zero component of b_1 and one of b_2, which exist. Set x_1 and x_2 to have value x_0 in these components and 0 otherwise. For these x_1 and x_2, (5.45) is a function of x_0 only and it is of the form (2.9), for which we know that equality cannot hold for all x_0.   QED

In the special case with $a_1 = a_2 = 0$, (5.45) corresponds, up to a multiplicative constant, to the product of two multivariate SN densities, both with non-null slant parameter. The implication is that this product cannot be expressed as some other multivariate SN density. By repeated application of this fact we can state the following: if we partition $Y \sim SN_d(0, \Omega, \alpha)$ in h blocks, so that $Y^\top = (Y_1^\top, \ldots, Y_h^\top)$, then joint independence of these h components requires that the parameters have a structure of the following form, in an obvious notation:

$$\Omega = \mathrm{diag}(\Omega_{11}, \ldots, \Omega_{hh}), \qquad \alpha = (0, \ldots, \alpha_j, \ldots, 0)^\top \tag{5.46}$$

so that the joint density (5.1) can be factorized into a product with separate variables.

This conclusion highlights an important aspect of the skew-normal distribution: independence among a set of components can hold only if at most one of them is marginally skew-normal. A direct implication of this fact is that two asymmetric marginal components of a multivariate skew-normal variate cannot be independent. Another implication is that the joint distribution of a set of independent skew-normal variables with non-zero slant (univariate or multivariate) cannot be multivariate SN.

As a further generalization, we now want to extend the above fact to a linear transformation $X = A^\top Y$, for a non-singular square matrix A.

Proposition 5.6 *Given* $Y \sim \mathrm{SN}_d(0, \Omega, \alpha)$, *consider the linear transform*

$$X = A^\top Y = \begin{pmatrix} X_1 \\ \vdots \\ X_h \end{pmatrix} = \begin{pmatrix} A_1^\top \\ \vdots \\ A_h^\top \end{pmatrix} Y \qquad (5.47)$$

where A is a $d \times d$ non-singular matrix and $(A_1, \ldots, A_h) = A$. Then $X_1, \ldots, X_h$ are mutually independent variables if and only if the following conditions hold simultaneously:

(a) $A_i^\top \Omega A_j = 0$ *for* $i \neq j$,
(b) $A_i^\top \Omega \omega^{-1} \alpha \neq 0$ *for at most one i.*

Proof When condition (a) holds, use of (5.12), (5.42) and (5.43) yields

$$\Omega_X = \mathrm{diag}(A_1^\top \Omega A_1, \ldots, A_h^\top \Omega A_h),$$

$$\alpha_X = \omega_X (A^\top \Omega A)^{-1} A^\top \Omega \omega^{-1} \alpha = \omega_X \begin{pmatrix} (A_1^\top \Omega A_1)^{-1} A_1^\top \Omega \omega^{-1} \alpha \\ \vdots \\ (A_h^\top \Omega A_h)^{-1} A_h^\top \Omega \omega^{-1} \alpha \end{pmatrix}.$$

From the last expression, it follows that, if condition (b) is fulfilled too, only one among the h blocks of α_X is non-zero. Hence the joint density of X can be factorized in an obvious manner and sufficiency is proved.

To prove necessity, note first that, if independence among $X_1, \ldots, X_h$ holds, then the joint density of X equals the product of the h marginal densities. Taking into account Proposition 5.5, equality can occur if only one block of α_X is not zero and Ω_X is block diagonal. QED

Corollary 5.7 *Given* $Y \sim \mathrm{SN}_d(0, \Omega, \alpha)$, *consider the partition* $\{s_1, \ldots, s_h\}$ *of* $\{1, \ldots, d\}$, *and let* $(Y_{s_1}^\top, \ldots, Y_{s_h}^\top)$ *denote the corresponding block partition of Y. Then $Y_{s_1}, \ldots, Y_{s_h}$ are mutually independent variables if and only if the following conditions hold simultaneously:*

(a) $\Omega_{s_i s_j} = 0$ *for* $i \neq j$,
(b) $\alpha_{s_i} \neq 0$ *for at most one i,*

where $\Omega_{s_i s_j}$ is the block portion of Ω formed by rows s_i and columns s_j.

The next result states that another classical property of the multivariate normal distribution holds for the SN case as well.

Proposition 5.8 *If* $Y \sim \mathrm{SN}_d(\xi, \Omega, \alpha)$, *its univariate components are pairwise independent if and only if they are mutually independent.*

Proof Necessity is trivial. To prove sufficiency, note firstly that closure under marginalization ensures that the joint distribution of any pair of marginal components Y_i and Y_j, say, is of type $SN_2(\xi', \Omega', \alpha')$, where the off-diagonal element of the matrix Ω' is Ω_{ij}. Also, from Proposition 5.6, Y_i and Y_j are independent if $\Omega_{ij} = 0$ and at least one between Y_i and Y_j is Gaussian, implying that the matrix Ω is diagonal and at least $d - 1$ univariate marginal components of Y are Gaussian, that is, $d - 1$ entries of δ defined in (5.11) should be zero. Mutual independence follows by noticing that the structure of the parameters Ω and α under pairwise independence guarantees that conditions (a) and (b) in Proposition 5.6 are fulfilled for $h = d$.   QED

The skew-normal distribution with 0 location shares with the normal family the distributional properties of the associated quadratic forms, because of the modulation invariance property of Proposition 1.4. More specifically, the connection is as follows.

Corollary 5.9 *If $Y \sim SN_d(0, \Omega, \alpha)$ and A is a $d \times d$ symmetric matrix, then*

$$Y^\top A Y \stackrel{d}{=} X^\top A X \tag{5.48}$$

where $X \sim N_d(0, \Omega)$.

This simple annotation immediately makes available the vast set of existing results for quadratic forms of multinormal variables. One statement of this type is property (5.7), among many others. The implications of modulation invariance are, however, not limited to a single quadratic form like in (5.48) by considering a q-valued even function $t(\cdot)$ in Proposition 1.4. For instance, the next result represents a form of Fisher–Cochran theorem.

Corollary 5.10 *If $Y \sim SN_d(0, I_d, \alpha)$ and $A_1, \ldots, A_n$ are symmetric positive semi-definite matrices with rank $r_1, \ldots, r_n$ such that $A_1 + \cdots + A_n = I_d$, then a necessary and sufficient condition that $Y^\top A_j Y \sim \chi^2_{r_j}$ and are independent is that $r_1 + \cdots + r_n = d$.*

5.1.7 A characterization result

A classical result of normal distribution theory is that, if all linear combinations $h^\top Z$ of a multivariate random variable Z have univariate normal distribution, then Z is multinormal. The next proposition states a matching fact for the normalized skew-normal distribution.

Proposition 5.11 *Consider a d-dimensional random variable Z such that $R = \mathbb{E}\{Z Z^\top\}$ is a finite and positive-definite correlation matrix. If, for any $h \in \mathbb{R}^d$ such that $h^\top R h = 1$, there exists a value α_h such that $h^\top Z \sim \mathrm{SN}(0, 1, \alpha_h)$, then $Z \sim \mathrm{SN}_d(0, R, \alpha)$ for some $\alpha \in \mathbb{R}^d$ and R is a correlation matrix.*

Proof Denote $T = h^\top Z \sim \mathrm{SN}(0, 1, \alpha_h)$ which has moment generating function $M_T(t) = 2\, e^{t^2/2}\Phi(\delta_h t)$, where δ_h is related to α_h as in (2.6). First, note that $b\, \delta_h = \mathbb{E}\{T\} = \mathbb{E}\{h^\top Z\} = h^\top \mu$, where $\mu = \mathbb{E}\{Z\}$ and $b = \sqrt{2/\pi}$. Therefore, choosing $h_0 = w^{-1} R^{-1}\mu$ where $w^2 = \mu^\top R^{-1}\mu$, so that $h_0^\top R h_0 = 1$, we obtain $b\delta_{h_0} = h_0^\top \mu = w$, which implies $b^2 > \mu^\top R^{-1}\mu$. Then, the vector

$$\alpha = (b^2 - w^2)^{-1/2} R^{-1}\mu = \left(b^2 - \mu^\top R^{-1}\mu\right)^{-1/2} R^{-1}\mu$$

exists and, after some simple algebra, it turns out to fulfil this equality:

$$\left(1 + \alpha^\top R\alpha\right)^{-1/2} \alpha^\top R h = \delta_h.$$

Hence, taking into account that $h^\top R h = 1$, we can write

$$M_T(t) = 2\, \exp\left(\tfrac{1}{2}t^2 h^\top R h\right) \Phi\left((1 + \alpha^\top R\alpha)^{-1/2} \alpha^\top R h\, t\right).$$

For any $u \in \mathbb{R}^d$, write it as $u = t h$ where t is a real and $h \in \mathbb{R}^d$ such that $h^\top R h = 1$. The moment generating function of Z at u is $M_Z(u) = \mathbb{E}\{\exp(t\, h^\top Z)\}$, which equals the above expression of $M_T(t)$ with $t h$ replaced by u. Comparing this with (5.10) evaluated at u, where δ is given by (5.11), we conclude that the moment generating function of Z is that of $\mathrm{SN}_d(0, R, \alpha)$, where R is a positive-definite correlation matrix. $\qquad$ QED

This characterization result could be used to develop the skew-normal distribution theory taking this property as the one which *defines* the probability distribution, following a similar route to that taken for the normal distribution, as recalled at the beginning of this section; see Rao (1973, Section 8a.1) and Mardia *et al.* (1979, Section 3.1.2).

5.1.8 Canonical form

We focus now on a specific type of linear transformation of a multivariate skew-normal variable, having special relevance for theoretical developments but to some extent also for practical reasons.

Proposition 5.12 *Given a variable $Y \sim \mathrm{SN}_d(\xi, \Omega, \alpha)$, there exists an affine transformation $Z^* = A_*(Y - \xi)$ such that $Z^* \sim \mathrm{SN}_d(0, I_d, \alpha_{Z^*})$, where $\alpha_{Z^*} = (\alpha_*, 0, \ldots, 0)^\top$ and α_* is defined by (5.37).*

Proof   Recall that in § 5.1 we have introduced the SN distribution assuming $\Omega > 0$ and the factorization $\Omega = \omega\bar{\Omega}\omega$ introduced right after (5.3); also let $\bar{\Omega} = C^{\mathsf{T}}C$ for some non-singular matrix C. If $\alpha \neq 0$, one can find an orthogonal matrix P with the first column proportional to $C\alpha$, while for $\alpha = 0$ we set $P = I_d$. Finally, define $A_* = (C^{-1}P)^{\mathsf{T}}\omega^{-1}$. It can be checked with the aid of formulae (5.41)–(5.43) for affine transformations that $Z^* = A_*(Y-\xi)$ has the stated distribution.   QED

The variable Z^*, which we shall sometimes refer to as a 'canonical variate', comprises d independent components. The joint density is given by the product of $d-1$ standard normal densities and at most one non-Gaussian component $\text{SN}(0, 1, \alpha_*)$; that is, the density of Z^* is

$$f_*(x) = 2 \prod_{i=1}^{d} \varphi(x_i)\,\Phi(\alpha_* x_1), \qquad x = (x_1, \ldots, x_d)^{\mathsf{T}} \in \mathbb{R}^d.$$

In § 5.1.5, α_* has emerged as the summary quantity which regulates the Mardia coefficients of multivariate skewness and kurtosis $\gamma_{1,d}^M$ and $\gamma_{2,d}^M$. Among the set of $\text{SN}(\xi, \Omega, \alpha)$ distributions sharing the same value of α_*, the canonical form can be regarded as the most 'pure' representative of this set, since all departure from normality is concentrated in a single component, independent from the others. Consequently, quantities which are invariant with respect to affine transformations can be computed more easily for the canonical form, and they hold for all distributions with the same value of α_*.

Therefore, expressions (5.39) of Mardia's measures could be derived as an instance of this scheme. A similar argument can be applied to compute the measures of multivariate skewness and kurtosis introduced by Malkovich and Afifi (1973). Since these measures are also invariant over affine transformations of the variable, we can reduce the problem to the canonical form, hence to the single univariate component possibly non-Gaussian. The implication is that we arrive again at expressions (5.39), equivalent to (5.34)–(5.35).

Inspection of the proof of Proposition 5.12 shows that, when $d > 2$, there exist several possible choices of A_*, hence many variables Z^*, all with the same distribution $f_*(x)$. However, this lack of uniqueness is not a problem. To draw an analogy, the canonical form plays a role loosely similar to the transformation which orthogonalizes the components of a multivariate normal variable, and also in that case the transformation is not unique.

Although Proposition 5.12 ensures that it is possible to obtain a canonical form, and we have remarked that in general there are many possible ways to do so, it is not obvious how to achieve the canonical form in practice. The next result explains this.

Proposition 5.13 *For $Y \sim \mathrm{SN}_d(\xi, \Omega, \alpha)$ define $M = \Omega^{-1/2}\Sigma\Omega^{-1/2}$, where $\Sigma = var\{Y\}$ and $\Omega^{1/2}$ is the unique positive definite symmetric square root of Ω. Let $Q\Lambda Q^{\top}$ denote a spectral decomposition of M, where without loss of generality we assume that the diagonal elements of Λ are arranged in increasing order, and $H = \Omega^{-1/2}Q$. Then*

$$Z^* = H^{\top}(Y - \xi)$$

has canonical form.

Proof From the assumptions made, it follows that $H^{-1} = Q^{\top}\Omega^{1/2}$ and $\Sigma = (H^{\top})^{-1}\Lambda H^{-1}$. In addition, use of (5.41) and (5.42) lends $\xi_{Z^*} = 0$ and $\Omega_{Z^*} = H^{\top}\Omega H = I_d$. In an obvious notation, therefore, we can write

$$\Sigma_{Z^*} = H^{\top}\Omega H - b^2\delta_{Z^*}\delta_{Z^*}^{\top} = I_d - b^2\delta_{Z^*}\delta_{Z^*}^{\top},$$

where $b^2 = 2/\pi$ and $\delta_{Z^*} = H^{\top}\omega\delta$ on recalling (5.32). Since we can also write

$$\Sigma_{Z^*} = H^{\top}\Sigma H = H^{\top}(H^{\top})^{-1}\Lambda H^{-1}H = \Lambda,$$

it follows that vector δ_{Z^*} can have at most one non-zero component, in the first position. This value will be $(\delta^{\top}\omega HH^{\top}\omega\delta)^{1/2} = (\delta^{\top}\bar{\Omega}^{-1}\delta)^{1/2} = \delta_*$, where the final equality follows from (5.38). Finally, from (5.12) and (2.15), we obtain

$$\alpha_{Z^*} = (1 - \delta_*)^{-1/2}(\delta_*, 0, \dots, 0)^{\top} = (\alpha_*, 0, \dots, 0)^{\top}. \qquad \text{QED}$$

So far we have employed the canonical form only to show simplified ways of computing multivariate coefficients of skewness and kurtosis. The next result, instead, seems difficult to prove without this notion. Recall that Proposition 5.1 implies that the multivariate SN density has a unique mode, like in the univariate case.

Proposition 5.14 *The unique mode of the distribution $\mathrm{SN}_d(\xi, \Omega, \alpha)$ is*

$$M_0 = \xi + \frac{m_0^*}{\alpha_*}\omega\bar{\Omega}\alpha = \xi + \frac{m_0^*}{\delta_*}\omega\delta, \qquad (5.49)$$

where δ and δ_* are given by (5.11) and by (5.38), respectively, and m_0^* is the mode of the univariate $\mathrm{SN}(0, 1, \alpha_*)$ distribution.

Proof Given a variable $Y \sim \mathrm{SN}_d(\xi, \Omega, \alpha)$, consider first the mode of the corresponding canonical variable $Z^* \sim \mathrm{SN}_d(0, I_d, \alpha_{Z^*})$. We find this mode by equating to zero the gradient of the density function, that is by solving the following equations with respect to $z_1, \ldots, z_d$:

$$z_1 \, \Phi(\alpha_* z_1) - \varphi_1(\alpha_* z_1)\, \alpha_* = 0, \qquad z_j \, \Phi(\alpha_* z_1) = 0 \quad \text{for } j = 2, \ldots, d \,.$$

The last $d - 1$ equations are fulfilled when $z_j = 0$, whilst the unique root of the first one corresponds to the mode, m_0^* say, of the $\mathrm{SN}(0, 1, \alpha_*)$ distribution. Therefore, the mode of Z^* is $M_0^* = (m_0^*, 0, \ldots, 0)^\top = (m_0^*/\alpha_*)\, \alpha_{Z^*}$. From Proposition 5.13, write $Y = \xi + \omega C^\top P Z^*$ and $\alpha_Z^* = P^\top C \alpha$. Since the mode is equivariant with respect to affine transformations, the mode of Y is

$$M_0 = \xi + \frac{m_0^*}{\alpha_*} \omega C^\top P P^\top C \alpha = \xi + \frac{m_0^*}{\alpha_*} \omega \bar{\Omega} \alpha = \xi + \frac{m_0^*}{\delta_*} \omega \delta,$$

where the last equality follows taking into account (5.11) and (5.38). QED

Equation (5.49) says that the mode lies on the direction of the vector $\omega \delta$ starting from location ξ. Recall from (5.31) that this is the same direction where the mean μ of this distribution is located. In other words, ξ, μ and M_0 are aligned points. Therefore, $\omega \delta$ is the direction where departure from Gaussianity displays more prominently its effect, and the intensity of this departure is summarized by α_*, or equivalently by δ_*. These conclusions are illustrated graphically in Figure 5.2, which refers to the case with

$$\xi = \begin{pmatrix} 3 \\ 5 \end{pmatrix}, \qquad \Omega = \begin{pmatrix} 2 & 2 \\ 2 & 4 \end{pmatrix}, \qquad \alpha = \begin{pmatrix} -5 \\ 2 \end{pmatrix}; \qquad (5.50)$$

the labels of the contour lines will be explained in Complement 5.2.

Besides the theoretical value of (5.49), there is also a practical one. Finding the mode of $\mathrm{SN}_d(\xi, \Omega, \alpha)$ requires a numerical maximization procedure, and in principle this search should take place in the d-dimensional Euclidean space, but by means of (5.49) we can restrict the search to a one-dimensional set, from ξ along the direction $\omega \delta$.

5.1.9 Bibliographic notes

Azzalini and Dalla Valle (1996) have introduced the multivariate version of the skew-normal distribution via the additive construction (5.19). Therefore, the parameterization adopted initially was $(\bar{\Psi}, \lambda)$, and the density function so obtained was written as a function of $\bar{\Omega}$ and α. However, at that stage the latter quantities did not yet appear to form a parameter set. They

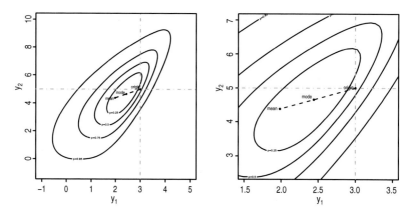

Figure 5.2 Contour lines plot of the bivariate skew-normal
density whose parameters are given in (5.50) with the mean value
and the mode superimposed, and their line of alignment. The
right-hand plot enlarges the central portion of the left-hand plot.

have also shown that the same family of distributions can be generated by
the conditioning mechanism (5.13), and have obtained some other results,
notably the chi-square property (5.7), the moment generating function and
the distribution function.

Azzalini and Capitanio (1999) have shown that the set of normalized
SN distributions could equally be parameterized by $(\bar{\Omega}, \alpha)$. From this basic
fact, much additional work has been developed, which represents the core
part of the exposition in the preceding pages. One of their results was the
canonical form, which has been explored further by Capitanio (2012); her
paper includes Propositions 5.13 and 5.14, and other results to be recalled
later. Before the general property of modulation invariance was discovered,
various specific instances were obtained; see for instance Loperfido (2001).
Loperfido (2010) verifies by a direct computation the coincidence of the
multivariate indices of Mardia with those of Malkovich and Afifi; in addition, he shows a direct correspondence between the canonical form and
the principal components, in the special case that α is an eigenvector of Ω.
Representation (5.25) via minima and maxima is an extension of a result
by Loperfido (2008). Proposition 5.8 seems to be new. Balakrishnan and
Scarpa (2012) have examined a range of other multivariate measures of
skewness for SN variates, some of vectorial type.

Following Azzalini and Capitanio (1999), most of the subsequent literature has adopted the (Ω, α) parameterization, or some variant of it, typically

(Ω, η) but often still denoted (Ω, α). Moreover in some papers, especially in the earlier ones, Ω denotes what here is $\bar{\Omega}$. So the reader should pay attention to which quantities are really intended. Adcock and Shutes (1999) adopt instead a parameterization of type (Ψ, λ), very similar to (5.30), which the author find preferable from the point of view of interpretability for financial applications. Their work is actually based on the extended distribution of § 5.3.

Genton *et al.* (2001) provide expression for moments up to the fourth order of SN variables and for lower-order moments of their quadratic forms. However, since what the authors denote Ω is a scale-free matrix, an adjustment is required to use these expressions in the general case: Ω must be interpreted as including the scale factor ω, that is, with the same meaning as in this book and δ must be replaced by $\omega\delta$ throughout.

The characterization result in Proposition 5.11 has been presented by Gupta and Huang (2002); the proof given here differs in two steps of the argument. Their paper includes other facts on linear and quadratic forms of skew-normal variates.

Javier and Gupta (2009) study the mutual information criterion for a multivariate SN distribution. Since its expression involves a quantity of type $\mathbb{E}\{\zeta_0(\alpha^\top Z)\}$, no explicit expression can be obtained, only reduced to a univariate integral, which is then expanded into an infinite series. Substantial additional work in this context, focusing on Shannon entropy and Kullback–Leibler divergence, has been carried out by Contreras-Reyes and Arellano-Valle (2012). Follow-up work by Arellano Valle *et al.* (2013) deals with similar issues for the broader class of skew-elliptical distributions, which are presented in Chapter 6.

Additional results on the multivariate SN distribution are recalled in the complements and in the set of problems at the end of the chapter.

5.2 Statistical aspects

5.2.1 *Log-likelihood function and parameter estimation*

Consider directly a regression setting where the ith component $y_i \in \mathbb{R}^d$ of $y = (y_1, \ldots, y_n)^\top$ is sampled from $Y_i \sim SN_d(\xi_i, \Omega, \alpha)$, with independence among the Y_i's. Assume that the location parameter ξ_i is related to a set of p explanatory variables x_i via

$$\xi_i^\top = x_i^\top \beta, \qquad i = 1, \ldots, n, \tag{5.51}$$

for some $p \times d$ matrix β of unknown parameters, where the covariates vector x_i has a 1 in the first position. We arrange the vectors $x_1, \ldots, x_n$ in a $n \times p$ matrix X (with $n > p$), which we assume to have rank p.

We commonly say that the DP is formed by (β, Ω, α), but duplicated elements must be removed; hence the more appropriate expression is

$$\theta^{\mathrm{DP}} = \begin{pmatrix} \mathrm{vec}(\beta) \\ \mathrm{vech}(\Omega) \\ \alpha \end{pmatrix}, \tag{5.52}$$

where $\mathrm{vec}(\cdot)$ is the operator which stacks the columns of a matrix and $\mathrm{vech}(\cdot)$ stacks the lower triangle, inclusive of the diagonal, of a symmetric matrix. From (5.3), the log-likelihood function is

$$\ell = c - \tfrac{1}{2} n \log \det(\Omega) - \tfrac{1}{2} n \operatorname{tr}(\Omega^{-1} S_\beta) + 1_n^\top \zeta_0(R_\beta \, \omega^{-1} \alpha) \tag{5.53}$$

where 1_n is the n-vector of all 1's,

$$c = -\tfrac{1}{2} n d \log(2\pi), \qquad R_\beta = y - X\beta, \qquad S_\beta = n^{-1} R_\beta^\top R_\beta,$$

ζ_0 is defined by (2.18). The notation $\zeta_0(x)$ when x is a vector must be interpreted as component-wise evaluation, similarly to (3.15); in the following, we shall employ the same convention also for other functions.

Maximization of this log-likelihood must be pursued numerically, over a parameter space of dimension $pd + d(d + 3)/2$, either by direct search of the function or using an EM-type algorithm. Here we describe a technique which works by direct optimization of the log-likelihood, combining analytical and numerical maximization.

First of all, notice that, for the purpose of this maximization, it is convenient to reparametrize temporarily the problem by replacing the component α of (5.52) with $\eta = \omega^{-1}\alpha$, since η enters only the final term of (5.53). Expression (5.53) without the last summand is the same as a Gaussian log-likelihood, and Ω does not enter the final term in the $(\mathrm{vec}(\beta), \mathrm{vech}(\Omega), \eta)$ parameterization. Using a well-known fact for Gaussian likelihoods, we can say immediately that, for any given β, maximization with respect to Ω is achieved at S_β. Plugging this expression into (5.53) lends the profile log-likelihood

$$\ell_*(\beta, \eta) = c - \tfrac{1}{2} n \log \det(S_\beta) - \tfrac{1}{2} n d + 1_n^\top \zeta_0(R_\beta \eta), \tag{5.54}$$

whose maximization must now be performed numerically with respect to $d(p + 1)$ parameter components. This process can be speeded up considerably if the partial derivatives

$$\frac{\partial \ell_*}{\partial \beta} = X^\top R_\beta S_\beta^{-1} - X^\top \zeta_1(R_\beta \eta) \, \eta^\top, \qquad \frac{\partial \ell_*}{\partial \eta} = R_\beta^\top \, \zeta_1(R_\beta \eta)$$

are supplied to a quasi-Newton algorithm. Once we have obtained the values $\hat{\beta}$ and $\hat{\eta}$ which maximize (5.54), the MLE of Ω is $\hat{\Omega} = S_{\hat{\beta}}$. From here we obtain $\hat{\omega}$, in an obvious notation, and the MLE of α as $\hat{\alpha} = \hat{\omega}\hat{\eta}$, recalling the equivariance property of MLE.

A form of penalized log-likelihood is possible, similarly to (3.30) with α^2 in (3.35) replaced by α_*^2. In this case, however, an equivalent of the profile log-likelihood (5.54) is not available.

After estimates of the parameters have been obtained, model adequacy can be examined graphically by comparing the fitted distributions with the data scatter, although in the multivariate case this must be reduced to a set of bivariate projections, or possibly trivariate projections when dynamic graphics can be employed.

Another device, aimed at an overall evaluation of the model fitting, is the perfect analogue of a diagnostic tool commonly in use for multivariate normal distributions (Healy, 1968), based on the empirical analogues of the Mahalanobis-type distances

$$d_i = (y_i - \hat{\xi}_i)^\top \hat{\Omega}^{-1}(y_i - \hat{\xi}_i), \qquad i = 1, \ldots, n, \qquad (5.55)$$

whose approximate reference distribution is χ_d^2, recalling (5.7). Here $\hat{\xi}_i^\top = x_i^\top \hat{\beta}$, the estimated location parameter for the ith observation, becomes a constant value $\hat{\xi}$ in the case of a simple sample. From these d_i's, we construct QQ-plots and PP-plots similar to those employed in the univariate case.

For a simple illustration of the above graphical devices, we make use of some variables of the Grignolino wine data. Specifically, we introduce the following multivariate response linear regression model:

$$(\text{tartaric\_acid, malic\_acid}) = \beta_0 + \beta_1 \, (\text{fixed\_acidity}) + \varepsilon,$$

where $\varepsilon \sim SN_2(0, \Omega, \alpha)$ and β_0, β_1 are vectors in $\mathbb{R}^2$.

After estimating β_0, β_1, Ω and α by maximum likelihood, the residuals of the fitted model have been plotted in Figure 5.3 with superimposed contour lines of the fitted error distribution. Each of these curves surrounds an area of approximate probability indicated by the respective curve label, using the method to be described in Complement 5.2. The visual impression is that the fitted distribution matches adequately the scatter of the residuals. It is true that there are four points out of 71 which fall outside the curve

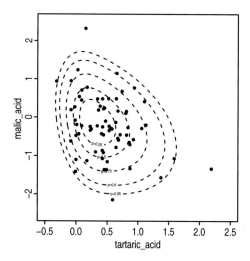

Figure 5.3 Grignolino wine data: empirical distribution of
residuals and fitted parametric model after the linear component
due to (fixed_acidity) has been removed from the joint distribution
of (tartaric_acid, malic_acid).

labelled $p = 0.95$, somewhat more than expected, but two of them are just
off the boundary. There is then some indication of a more elongated 'tail'
than the normal one, but only in a mild form.

The overall impression of an essentially adequate data fit is supported
also by the QQ-plot based on the distances (5.55), displayed in the left
panel of Figure 5.4; there is only one point markedly distant from the ideal
identity line. The right panel of the same figure displays the corresponding PP-plot, and compares it with the similar construct under normality
assumption and least-squares (LS) fit. There is a quite clear indication of
an improvement provided by the SN fit, whose points are visibly closer to
the identity lines than the LS points.

Bibliographic notes

The above-described technique for maximization of the log-likelihood and
the Healy-type graphical diagnostics have been put forward by Azzalini
and Capitanio (1999, Section 6.1). Using these expressions of the partial
derivatives, Azzalini and Genton (2008) deduce that the profile log-likelihood always has a stationary point at the point where β equals the least-
squares estimate and $\eta = 0 = \alpha$; they also extend the result to a broader
setting where the G_0 distribution of the modulation factor is not necessarily

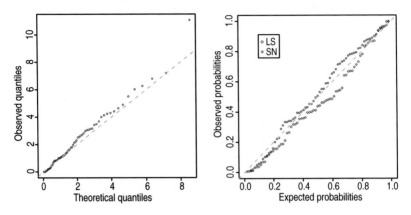

Figure 5.4 Grignolino wine data: QQ-plot (left) and PP-plot (right) of the same fitting as Figure 5.3; in the right plot the points corresponding to the least-squares fit are also displayed.

Gaussian. Instead of employing graphical diagnostics, the distributional assumption may be examined using a formal test procedure using the method proposed by Meintanis and Hlávka (2010) based on the empirical moment generating function; another option is to use the general procedure of Jiménez-Gamero *et al.* (2009) based the empirical characteristic function.

5.2.2 Fisher information matrix

Computation of the information matrix associated with the log-likelihood (5.53) is technically intricate. We only summarize the main facts, referring the reader to Arellano-Valle and Azzalini (2008) for a full treatment.

For mathematical convenience we consider again the parameterization, now denoted θ^{SP}, which replaces α in (5.52) with $\eta = \omega^{-1}\alpha$. Define the $d^2 \times [d(d+1)/2]$ duplication matrix D_d such that $\mathrm{vec}(M) = D_d\,\mathrm{vech}(M)$ for a symmetric matrix M, and let $Z_2 = -\mathrm{diag}(\zeta_2(R_\beta\eta)) > 0$. Then it can be shown that

$$
-\frac{\partial^2 \ell(\theta^{\mathrm{SP}})}{\partial \theta^{\mathrm{SP}}\,\partial(\theta^{\mathrm{SP}})^{\top}}
$$
$$
= \begin{pmatrix}
\Omega^{-1} \otimes (X^{\top}X) + (\eta\eta^{\top}) \otimes (X^{\top}Z_2X) & \cdot & \cdot \\
D_d^{\top}[\Omega^{-1} \otimes (\Omega^{-1}R_\beta^{\top}X)] & \tfrac{1}{2}n\,D_d^{\top}(\Omega^{-1} \otimes V)D_d & \cdot \\
I_d \otimes u^{\top} - \eta \otimes U^{\top} & 0 & R_\beta^{\top}Z_2R_\beta
\end{pmatrix}
$$
$$
(5.56)
$$

where $u = X^{\top}\zeta_1(R_\beta\eta)$, $U = X^{\top}Z_2 R_\beta$, $V = \Omega^{-1}(2S_\beta - \Omega)\Omega^{-1}$ and the upper triangle must be filled symmetrically. Evaluation of this matrix at the MLE, $\hat{\theta}^{\mathrm{SP}}$, gives the observed information matrix, $\mathcal{J}(\hat{\theta}^{\mathrm{SP}})$.

To compute the expected value of (5.56), consider $U \sim \mathrm{N}(0, \bar{\alpha}^2)$, where $\bar{\alpha}^2 = \alpha_*^2/(1 + 2\alpha_*^2)$, and define

$$a_0 = \mathbb{E}\left\{\Phi(U)^{-1}\right\}, \qquad a_1 = \mathbb{E}\left\{\Phi(U)^{-1}U\mu_c\right\},$$
$$A_2 = \mathbb{E}\left\{\Phi(U)^{-1}\left(U^2\mu_c\mu_c^\top + \Omega_c\right)\right\},$$

where $\mu_c = \alpha_*^{-2}\Omega\eta = (\eta^\top\Omega\eta)^{-2}\Omega\eta$ and $\Omega_c = \Omega - \alpha_*^{-2}\Omega\eta\eta^\top\Omega$. Evaluation of these coefficients requires numerical integration, but only in the limited form of three 1-dimensional integrals, irrespective of d. The expected Fisher information matrix for θ^{SP} is then

$$\mathcal{I}(\theta^{\mathrm{SP}}) = \begin{pmatrix} (\Omega^{-1} + c_2 a_0 \eta\eta^\top) \otimes (X^\top X) & \cdot & \cdot \\ c_1 D_d^\top[\Omega^{-1} \otimes (\eta 1_n^\top X)] & \frac{1}{2}n D_d^\top(\Omega^{-1} \otimes \Omega^{-1})D_d & \cdot \\ A_1^\top \otimes (1_n^\top X) & 0 & n c_2 A_2 \end{pmatrix},$$
$$\tag{5.57}$$

where

$$c_k = (1 + k\,\eta^\top\Omega\eta)^{-1/2}\,2\,(b/2)^k, \qquad A_1 = c_1(I_d + \eta\,\eta^\top\Omega)^{-1} - c_2\eta a_1^\top.$$

Conversion of either type of information matrix, $\mathcal{J}(\theta^{\mathrm{SP}})$ or $\mathcal{I}(\theta^{\mathrm{SP}})$, to its counterpart for θ^{DP} requires the Jacobian matrix of the partial derivatives of θ^{SP} with respect to θ^{DP}, that is,

$$D_{\theta^{\mathrm{DP}}}(\theta^{\mathrm{SP}}) = \begin{pmatrix} I_{pd} & 0 & 0 \\ 0 & I_{d(d+1)/2} & 0 \\ 0 & D_{32} & \omega^{-1} \end{pmatrix}$$

where

$$D_{32} = -\frac{1}{2}\sum_{i=1}^{d}(\Omega_{ii})^{-3/2}(\alpha^\top E_{ii} \otimes E_{ii})D_d,$$

having denoted by E_{ii} the $d \times d$ matrix having 1 in the (i, i)th entry and 0 otherwise. Then the expected information matrix for θ^{DP} is

$$\mathcal{I}^{\mathrm{DP}}(\theta^{\mathrm{DP}}) = D_{\theta^{\mathrm{DP}}}(\theta)^\top \mathcal{I}(\theta^{\mathrm{SP}})\, D_{\theta^{\mathrm{DP}}}(\theta^{\mathrm{SP}}) \tag{5.58}$$

and a similar expression holds for $\mathcal{J}(\theta^{\mathrm{DP}})$.

5.2.3 The centred parameterization

We examine the multivariate extension of the centred parameterization discussed in §3.1.4 for the univariate case. For simplicity of exposition, we refer to the case $p = 1$, hence with a constant location parameter ξ. This does not represent a restriction, since (3.23) indicates that only the first regression component changes moving between DP and CP.

A direct extension of the CP notion to the multivariate case is represented by (μ, Σ, γ_1), where the first two components are given by (5.31) and (5.32), respectively, and γ_1 is the d-vector of marginal coefficients of skewness obtained by component-wise application of (2.28) to each component of (5.11), via (2.26). Formally, only the non-replicated entries of Σ, namely vech(Σ), enter the parameter vector.

The above description says how to obtain CP from DP. As DP spans its feasible range, which is only restricted by the condition $\Omega > 0$, CP spans a corresponding set. An important difference is that, while the DP components are variation-independent, the same is not true for the CP components, if $d > 1$. However, if a certain parameter combination (μ, Σ, γ_1) belongs to the feasible CP parameter set, then there is a unique inverse point in the DP space. To see this, from the jth component of γ_1 obtain $\mu_{z,j}$ inverting (2.28) and from here δ_j, for $j = 1, \ldots, d$. This gives a d-vector δ and a diagonal matrix σ_z whose jth non-zero entry is obtained from δ_j using the second expression of (2.26). After forming the diagonal matrix σ with the square-root of the diagonal entries of Σ, the first two DP components are given by

$$\xi = \mu + \sigma\sigma_z^{-1}\mu_z, \qquad \omega = \sigma\sigma_z^{-1}, \qquad \Omega = \Sigma + \omega\mu_z\mu_z^\top\omega$$

and α is as in (5.12). Therefore the CP, (μ, Σ, γ_1), represents a legitimate parameterization of the multivariate SN family.

One of the arguments in support of the CP in the univariate case was the simpler interpretation of mean, standard deviation and coefficient of skewness, compared to the corresponding component of the DP. This aspect holds *a fortiori* in the multivariate context, where the values taken on by the components of α are not easily interpretable. As a numerical illustration of this point, consider two sets of parameters, $(\Omega, \alpha^{(1)})$ and $(\Omega, \alpha^{(2)})$, where

$$\Omega = \begin{pmatrix} 2 & 1 & 3 \\ 1 & 2 & 4 \\ 3 & 4 & 9 \end{pmatrix}, \qquad \alpha^{(1)} = \begin{pmatrix} 5 \\ -3 \\ 4 \end{pmatrix}, \qquad \alpha^{(2)} = \begin{pmatrix} 5 \\ -3 \\ -4 \end{pmatrix}$$

whose corresponding coefficients of marginal skewness, rounded to two decimal digits, are

$$\gamma_1^{(1)} = (0.85, \ 0.04, \ 0.16)^\top, \qquad \gamma_1^{(2)} = (0.00, \ -0.21, \ -0.07)^\top,$$

respectively. Visibly, consideration of an individual component of α does not provide information on the corresponding component of γ_1, in fact not even on its sign, while in the univariate case there at least exists a monotonic relationship between α and γ_1.

For inferential purposes, ML estimates of the CP are simply obtained by transformation to the CP space of the ML estimates of the DP, by the equivariance property. The Fisher CP information matrix, $I(\theta^{cp})$, is obtained from (5.57) by a transformation similar to (5.58) where the Jacobian matrix is now constituted by the partial derivatives of θ^{sp} with respect to θ^{cp}, which is also given by Arellano-Valle and Azzalini (2008). For mathematical convenience, an intermediate parameterization between θ^{sp} and θ^{cp} is introduced; consequently, this Jacobian matrix is expressed as the product of two such matrices.

Arellano-Valle and Azzalini (2008) have further considered the asymptotic behaviour of the resulting information matrix in the limiting case where $\gamma_1 \to 0$, or equivalently $\alpha \to 0$. While a limiting form of $I(\theta^{cp})$ has been stated in the quoted paper, subsequent analysis has raised doubts on the correctness of this result, specifically on the diagonal block pertaining to γ_1, when $d > 1$. Further investigation on this issue is therefore required. If $d = 1$ the asymptotic expression is in agreement with the results of Chapter 3.

The previous passage prevents, at least currently, making use of the multivariate CP for inferential purposes in a neighbourhood of $\gamma_1 = 0$. Still, we feel like considering the usage of the CP in situations separate from $\gamma_1 = 0$, because the problematic aspects at one point do not prevent their use over the remaining parameter space, taking into account considerations on interpretability of the parameters discussed earlier.

5.3 Multivariate extended skew-normal distribution

5.3.1 Definition and basic properties

A d-dimensional version of the extended skew-normal distribution examined in §2.2 is given by

$$\varphi_d(x; \bar{\Omega}, \alpha, \tau) = \varphi_d(x; \bar{\Omega}) \frac{\Phi\{\alpha_0 + \alpha^\top x\}}{\Phi(\tau)}, \qquad x \in \mathbb{R}^d, \tag{5.59}$$

where $\tau \in \mathbb{R}$,

$$\alpha_0 = \tau(1 + \alpha^\top \bar{\Omega}\alpha)^{1/2} \tag{5.60}$$

and the other terms are as in (5.1). Using Lemma 5.2, it is straightforward to confirm that (5.59) integrates to 1. Similarly to the univariate case, τ effectively vanishes when $\alpha = 0$. A slightly different parameterization in use regards α_0 as a parameter component in place of τ, while here we shall use α_0 only as a short-hand notation for (5.60).

If Z has density (5.59) and $Y = \xi + \omega Z$ as in (5.2), the density of Y at $x \in \mathbb{R}^d$ is

$$\varphi_d(x - \xi; \Omega)\, \Phi\left\{\alpha_0 + \alpha^\top \omega^{-1}(x - \xi)\right\}\, \Phi(\tau)^{-1} \qquad (5.61)$$

with the same notation of (5.3). In this case, we write $Y \sim \mathrm{SN}_d(\xi, \Omega, \alpha, \tau)$, where again the presence of the fourth parameter component indicates that the distribution is 'extended'.

Using Lemma 5.3, the moment generating function of the distribution $Y \sim \mathrm{SN}_d(\xi, \Omega, \alpha, \tau)$ is readily seen to be

$$M(t) = \exp(t^\top \xi + \tfrac{1}{2} t^\top \Omega t)\, \Phi(\tau + \delta^\top \omega t)\, \Phi(\tau)^{-1}, \qquad t \in \mathbb{R}^d, \qquad (5.62)$$

where δ is as in (5.11).

From $M(t)$, which matches closely (5.10) of the SN case, we can derive the distribution for marginal block components and for affine transformations of Y. Specifically, if Y is partitioned as $Y^\top = (Y_1^\top, Y_2^\top)$ where the two blocks have size h and $d-h$, as in § 5.1.4, then marginally

$$Y_1 \sim \mathrm{SN}_h(\xi_1, \Omega_{11}, \alpha_{1(2)}, \tau), \qquad (5.63)$$

where the first three parameter components are the same as the SN case given by (5.29). For an affine transformation $X = c + A^\top Y$, where A is a full-rank $d \times h$ matrix ($h \le d$) and $c \in \mathbb{R}^h$, we have

$$X \sim \mathrm{SN}_h(\xi_X, \Omega_X, \alpha_X, \tau),$$

where the first three parameter components are given by (5.41)–(5.43).

Similarly to its univariate counterpart, density (5.59) does not satisfy the conditions for the property of modulation invariance (1.12). Hence the results of § 5.1.6 on quadratic forms of SN variates do not carry on here.

A mathematically appealing aspect of this distribution is first suggested by the observation that, if $X \sim \mathrm{SN}_d(\xi, \Omega, \alpha)$, then the conditional density of X given that a subset of its components takes on a certain value is of type (5.61); see Problem 5.12. This property is a simplified version of the closure property of the next paragraph.

5.3.2 Conditional distribution and conditional independence

An important property of the family (5.61) is its closure with respect to conditioning on the values taken on by some components. To see this, partition $Y \sim \mathrm{SN}_d(\xi, \Omega, \alpha, \tau)$ as $Y = (Y_1^\top, Y_2^\top)^\top$, where Y_1 has dimension h, and examine the conditional distribution of Y_2 given that $Y_1 = y_1$. Recall that, if Y was a $N_d(\xi, \Omega)$ variable, the parameters of the conditional distribution would be

$$\xi_{2\cdot 1} = \xi_2 + \Omega_{21}\Omega_{11}^{-1}(y_1 - \xi_1), \qquad \Omega_{22\cdot 1} = \Omega_{22} - \Omega_{21}\Omega_{11}^{-1}\Omega_{12} \qquad (5.64)$$

and these quantities emerge again when we take the ratio of the normal densities involved by $(Y_2|Y_1 = y_1)$. Then, using (5.63), the conditional density of Y_2 given $Y_1 = y_1$ is

$$\varphi_{d-h}(y_2 - \xi_{2\cdot 1}; \Omega_{22\cdot 1}) \frac{\Phi\left\{\alpha_0' + \alpha_2^\top \omega_2^{-1}(y_2 - \xi_{2\cdot 1})\right\}}{\Phi(\tau_{2\cdot 1})}, \qquad y_2 \in \mathbb{R}^{d-h}, \qquad (5.65)$$

where

$$\begin{aligned}
\tau_{2\cdot 1} &= \tau \left(1 + \alpha_{1(2)}^\top \bar{\Omega}_{11} \alpha_{1(2)}\right)^{1/2} + \alpha_{1(2)}^\top \omega_1^{-1}(y_1 - \xi_1), \\
\alpha_0' &= \tau_{2\cdot 1}(1 + \alpha_{2\cdot 1}^\top \bar{\Omega}_{22\cdot 1}^{-1} \alpha_{2\cdot 1})^{1/2}, \\
\alpha_{2\cdot 1} &= \omega_{22\cdot 1}\omega_2^{-1}\alpha_2, \\
\omega_{22\cdot 1} &= (\Omega_{22\cdot 1} \odot I_{d-h})^{1/2}
\end{aligned} \qquad (5.66)$$

and we have used the notation in (5.27) and (5.28) on p. 130. To conclude, write

$$(Y_2|Y_1 = y_1) \sim SN_{d-h}(\xi_{2\cdot 1}, \Omega_{22\cdot 1}, \alpha_{2\cdot 1}, \tau_{2\cdot 1}) \qquad (5.67)$$

which states the property of closure with respect to conditioning.

The above expression of $\alpha_{2\cdot 1}$ provides the key to interpret the presence of null components of α. Since $\alpha_{2\cdot 1} = 0$ if and only if $\alpha_2 = 0$, then $\alpha_2 = 0$ means that $(Y_2|Y_1 = y_1)$ is Gaussian. Consequently, when the rth component of α is null, the conditional distribution of Y_r given all other components is Gaussian. These facts hold both in the ESN and in the SN case, since (5.67) holds also when $\tau = 0$, with a simplification in $\tau_{2\cdot 1}$.

This type of argument can be carried on to examine conditional independence among components of the distribution of $(Y_2|Y_1 = y_1)$. Specifically, bearing in mind the relationship between $\alpha_{2\cdot 1}$ and α_2 as given in (5.66) and that $\Omega_{22\cdot 1} = (\Omega^{-1})_{22}$, conditions for conditional independence can be stated directly as conditions on α and Ω^{-1}. This fact is exploited to obtain the next result.

Proposition 5.15 *Consider any three-block partition $Y = (Y_1^\top, Y_{2a}^\top, Y_{2b}^\top)^\top$ of $Y \sim SN_d(\xi, \Omega, \alpha, \tau)$. Then Y_{2a} and Y_{2b} are conditionally independent given Y_1 if and only if the following conditions hold simultaneously:*

(a) $(\Omega^{-1})_{ab} = 0$,
(b) at least one of α_a and α_b is the null vector,

where α_a and α_b denote the subsets of α associated with Y_{2a} and Y_{2b}, respectively, and $(\Omega^{-1})_{ab}$ is the corresponding block portion of Ω^{-1}.

Proof   Since the value of τ does not affect the conditional independence among the components of $Y_2 = (Y_{2a}^\top, Y_{2b}^\top)^\top$, we can argue as if $\tau = 0$. Then the statement can be proved recalling that independence requires that the parameters of the conditional distribution must have the structure as in (5.46). In the present case, that structure holds for $h = 2$, the pertaining scale matrix is $\Omega_{22\cdot 1}$, that is, the scale matrix of the conditional distribution given Y_1, and the slant parameter $\alpha_{2\cdot 1}$ is computed from (5.66).   QED

The property of closure under conditioning and the last proposition form the basis for developing graphical models of ESN variables. Some results in this direction will be presented in § 5.3.5.

5.3.3 Stochastic representations and distribution function

Some stochastic representations of the multivariate SN distribution extend naturally to the ESN case; others do not, or at least no such extension is known at the time of writing.

A stochastic representation via a conditioning mechanism is as follows. Starting from (X_0, X_1) distributed as in (5.14), a standard computation says that, for any $\tau \in \mathbb{R}$,

$$Z = (X_0 | X_1 + \tau > 0) \sim \mathrm{SN}_d(0, \bar{\Omega}, \alpha(\delta), \tau) \qquad (5.68)$$

where $\alpha(\delta)$ is given by (5.12), similarly to the first expression in (5.15).

Representation (5.68) indicates how to compute the distribution function of Z. By a computation similar to (2.48), write

$$
\begin{aligned}
\mathbb{P}\{Z \le z\} &= \mathbb{P}\{X_0 \le z | X_1 + \tau > 0\} \\
&= \mathbb{P}\{(X_0 \le z) \cap (-X_1 < \tau)\} / \mathbb{P}\{-X_1 < \tau\} \\
&= \Phi_{d+1}((z^\top, \tau)^\top; \tilde{\Omega}) / \Phi(\tau), \qquad (5.69)
\end{aligned}
$$

where $\tilde{\Omega}$ is a matrix similar to Ω^* in (5.14) with δ replaced by $-\delta$. The general case $\mathrm{SN}_d(\xi, \Omega, \alpha, \tau)$ is handled as usual by reduction to a normalized variable Z. Therefore, the distribution function of a d-dimensional ESN, and then also of an SN, variable is computed by evaluating a suitable $(d+1)$-dimensional normal distribution function.

To introduce a form of additive representation of an ESN variate, start from the independent variables $U_0 \sim \mathrm{N}_d(0, \bar{\Psi})$, where $\bar{\Psi}$ is a full-rank correlation matrix, and $U_{1,-\tau}$ which is a $\mathrm{N}(0, 1)$ variable truncated below $-\tau$

for some $\tau \in \mathbb{R}$. Then a direct extension of (2.43), using the notation of (5.18)–(5.19), is

$$Z = D_\delta \, U_0 + \delta \, U_{1,-\tau} \tag{5.70}$$

such that $Z \sim \mathrm{SN}_d(0, \bar{\Omega}, \alpha, \tau)$ where $\bar{\Omega}$ and α are related to $\bar{\Psi}$ and δ as in (5.20)–(5.22); see Problem 5.13.

For the reasons discussed in §2.2.2 for the univariate case, representation (5.70) is more convenient than (5.68) for random number generation.

5.3.4 Cumulants and related quantities

From (5.62) the cumulant generating function of $Y \sim \mathrm{SN}_d(\xi, \Omega, \alpha, \tau)$ is

$$K(t) = \log M(t) = \xi^\top t + \tfrac{1}{2} t^\top \Omega t + \zeta_0(\tau + \delta^\top \omega t) - \zeta_0(\tau), \qquad t \in \mathbb{R}^d,$$

where $\zeta_0(x)$ is defined by (2.18) along with its successive derivatives $\zeta_r(x)$. Evaluation at $t = 0$ of the first two derivatives of $K(t)$ leads to

$$\mathbb{E}\{Y\} = \xi + \zeta_1(\tau) \, \omega \, \delta = \xi + \omega \mu_z, \tag{5.71}$$

$$\mathrm{var}\{Y\} = \Omega + \zeta_2(\tau) \, \omega \, \delta \, \delta^\top \omega = \omega \, \Sigma_z \, \omega, \tag{5.72}$$

where

$$\mu_z = \mathbb{E}\{Z\} = \zeta_1(\tau) \, \delta, \qquad \Sigma_z = \mathrm{var}\{Z\} = \bar{\Omega} + \zeta_2(\tau) \, \delta \, \delta^\top$$

refer to $Z \sim \mathrm{SN}_d(0, \bar{\Omega}, \alpha, \tau)$. Higher-order derivatives of $K(t)$ are

$$\frac{\mathrm{d}^r}{\mathrm{d}t_i \, \mathrm{d}t_j \, \cdots \, \mathrm{d}t_h} K(t) = \zeta_r(\tau + \delta^\top \omega \, t) \, \omega_i \, \omega_j \cdots \omega_h \, \delta_i \, \delta_j \cdots \delta_h. \tag{5.73}$$

Proceeding similarly to §5.1.5, we obtain that the Mardia coefficients of multivariate skewness and kurtosis are

$$\gamma_{1,d}^M = \left(\frac{\zeta_3(\tau)}{\zeta_1(\tau)^3} \right)^2 \left(\mu_z^\top \Sigma_z^{-1} \mu_z \right)^3 = \zeta_3(\tau)^2 \left(\frac{\delta_*^2}{1 + \zeta_2(\tau) \delta_*^2} \right)^3, \tag{5.74}$$

$$\gamma_{2,d}^M = \frac{\zeta_4(\tau)}{\zeta_1(\tau)^4} \left(\mu_z^\top \Sigma_z^{-1} \mu_z \right)^2 = \zeta_4(\tau) \left(\frac{\delta_*^2}{1 + \zeta_2(\tau) \delta_*^2} \right)^2, \tag{5.75}$$

where δ_* is as in (5.38). The two final expressions match those in (2.46) and (2.47) evaluated at δ_*, except that the Mardia coefficient of skewness when $d = 1$ corresponds to the square of the univariate coefficient. Therefore the range of $(\gamma_{1,d}^M, \gamma_{2,d}^M)$ is the same as pictured in Figure 2.5 provided the γ_1-axis is square transformed.

5.3.5 Conditional independence graphs

The aim of this section is to present some introductory notions on *graphical models* for ESN variables, specifically in the form of conditional independence graphs. For background material on graphical models, we refer the reader to the monographs of Cox and Wermuth (1996) and Lauritzen (1996).

A graphical model is constituted by a graph, denoted $\mathscr{G} = (V, E)$, where the set V of the vertices or nodes is formed by the components of a multivariate random variable $Y = (Y_1, \ldots, Y_d)^\top$ and the set E of edges connecting elements of V is chosen to represent the dependence structure induced by the distribution of Y.

A conditional independence graph is a construction with the additional requirements that (a) the graph is *undirected*, which means that an edge is a set of two unordered elements of V, so that we do not make distinction among (i, j), (j, i) and $\{i, j\}$, and (b) the nodes i and j are not connected if Y_i and Y_j are conditionally independent given all other components of Y, for $i \neq j$. Formally we write $\{i, j\} \notin E$ if $Y_i \perp\!\!\!\perp Y_j|$(all other variables), where the symbol $\perp\!\!\!\perp$ denotes independence.

We now explore the above concepts when Y has a multivariate ESN distribution. The focus is on this family because closure with respect to conditioning plays a fundamental role here. From Proposition 5.15 it is immediate to state the following result.

Corollary 5.16 (Pairwise conditional independence)   *If* $Y = (Y_1, \ldots, Y_d)^\top$ $\sim \mathrm{SN}_d(\xi, \Omega, \alpha, \tau)$, *then*

$$Y_i \perp\!\!\!\perp Y_j | \textit{(all other variables)}$$

if and only if the following conditions hold simultaneously:

(a) $\Omega^{ij} = 0$,
(b) $\alpha_i \alpha_j = 0$,

where Ω^{ij} *denotes the* (i, j)*th entry of* Ω^{-1}.

This statement lends the operational rule to specify the conditional independence graph associated with Y:

$$(i, j) \in E \quad \Longleftrightarrow \quad \{\Omega^{ij} \neq 0 \quad \text{or} \quad \alpha_i \alpha_j \neq 0\}. \tag{5.76}$$

When $\alpha = 0$, we recover the classical rule for the Gaussian case based solely on the elements of the concentration matrix, that is, the inverse of the variance matrix.

So far, the graph built via (5.76) reflects the conditional independence for a pair of variables, but we are interested in establishing all conditional independence statements implied by this structure. This extension is possible thanks to the *global Markov property*, which applies to continuous random variables with density positive everywhere on the support, such as the ESN family; see the monographs cited earlier for a detailed discussion of these aspects. In essence, the global Markov property can be described as follows: if A, B and C are disjoint subsets of vertices and C separates A from B, then conditional independence $Y_A \perp\!\!\!\perp Y_B | Y_C$ holds for the corresponding set of variables, Y_A, Y_B, Y_C. Recall that C *separates A from B* if there is no sequence of edges connecting a node in A with a node in B without going through some node in C.

Clearly, for a given pair (Ω, α), the corresponding conditional independence graph is uniquely specified by (5.76). The converse is not true: a given graph is compatible with several patterns of (Ω, α). For instance a *complete* graph, where an edge exists between any pair of distinct vertices, can be obtained both from the pair $(I_d, a \, 1_d)$ where $a \neq 0$ and from the pair (Ω, α) where Ω^{-1} has no zero entries and α is arbitrary.

Stochastic representation (5.68) of Y indicates how this variable is related to a suitable $(d+1)$-dimensional normal variable X, as specified in (5.14). The next proposition indicates how the respective conditional independence graphs are related.

Proposition 5.17 *Given the conditional independence graph $\mathscr{G}_X$ of X with distribution (5.14), the conditional independence graph $\mathscr{G}_Z$ of Z, defined in (5.68), is uniquely identified and can be obtained by adding those edges needed to make the boundary of the vertex associated with X_1 complete and by deleting this vertex and the corresponding edges.*

Proof By making use of (5.12), the concentration matrix of $X = (X_0^\top, X_1)^\top$ can be written as

$$(\Omega^*)^{-1} = \begin{pmatrix} A & -\alpha c \\ -\alpha^\top c & c^2 \end{pmatrix}, \tag{5.77}$$

where $A = \bar{\Omega}^{-1} + \alpha\alpha^\top$ and $c = (1 - \delta_*^2)^{-1/2} > 0$ with δ_* defined by (5.38). If $A_{ij} \neq 0$, so that the edge (i, j) exists in $\mathscr{G}_X$, then this edge will also exist in $\mathscr{G}_Z$, from (5.76). If $A_{ij} = 0$ and $\alpha_i\alpha_j \neq 0$, then $\bar{\Omega}^{ij} \neq 0$. Hence we must add an edge (i, j) if vertex X_1 is connected to both i and j. QED

For simplicity of notation, Proposition 5.17 has been stated for the case of a normalized variable Z with zero location and unit scale factors, but it holds for the general case as well.

We now examine the conditions for separation when Y has an ESN distribution. In this case the possible presence of nodes with marginal Gaussian distribution introduces constraints on the structure of conditional dependence, so that some patterns are inhibited. Moreover, the existence of Gaussian nodes may provide an indication of the presence of 0 elements in α. To distinguish the two types of nodes, we mark the nodes having Gaussian marginal distribution with 'G', and the others with 'SN', dropping the 'E' of ESN for mere simplicity of notation. Correspondingly, V is partitioned into two disjoint sets, V_G and V_{SN}. The boundary set of vertex i formed by all vertices which share an edge with i is denoted by bd(i).

Proposition 5.18 *Consider the three-block partition $Y^\top = (Y_A^\top, Y_B^\top, Y_C^\top)$ where A, B and C are disjoint subsets of indices and $Y \sim SN_d(\xi, \Omega, \alpha, \tau)$. If C separates A from B, one among the three following conditions must hold:*

(a) $A \cup C \subseteq V_G$,

(b) $B \cup C \subseteq V_G$,

(c) $C \nsubseteq V_G$.

Proof   Recall Corollary 5.7 which clearly holds also for ESN distributions. Since Y obeys the global Markov property, the fact that C separates A from B corresponds to the independence relationship $Y_A \perp\!\!\!\perp Y_B | Y_C$. Then Corollary 5.7 implies that $\bar\Omega^{AB} = 0$ and at least one of α_A and α_B is the null vector. Therefore, from (5.12), at least one of the two following equalities must hold:

$$\alpha_A = k\,(\bar\Omega^{AA}\delta_A + \bar\Omega^{AC}\delta_C) = 0, \qquad \alpha_B = k\,(\bar\Omega^{BB}\delta_B + \bar\Omega^{BC}\delta_C) = 0,$$

in an obvious notation, for some $k > 0$. Conditions (a) and (b) then follow because both $\bar\Omega^{AA} > 0$ and $\bar\Omega^{BB} > 0$. If both (a) and (b) fail, separation can only occur under (c).   QED

Corollary 5.19   *Let (A, B, C) be a partition of V such that $A \cup C \subseteq V_G$. If C separates A from B, then $\alpha_A = 0$.*

Proposition 5.20   *If $i \in V_G$ and bd(i) $\cap V_{SN} = \{h\}$ [i.e., bd(i) has only one vertex in V_{SN}], then $\alpha_i \neq 0$.*

Proof   Let h be the unique non-Gaussian vertex in bd(i). Then from (5.12) we have $\alpha_i \propto \Omega^{ih}\delta_h$. Since $\delta_h \neq 0$, it follows that $\alpha_i = 0$ if and only if $\Omega^{ih} = 0$, implying $(i, h) \notin E$.   QED

Corollary 5.21   *If $i, j \subseteq V_G$ and both* $\mathrm{bd}(i)$ *and* $\mathrm{bd}(j)$ *have exactly one vertex in* V_{SN}, *then* $(i, j) \in E$.

Proof   Immediate from Propositions 5.18 and 5.20.   QED

Operationally, these statements allow us to define two rules for checking the admissibility of a marked graph, with vertices labelled G or SN: (a) in any three-set partition of a marked graph, a subset of G vertices cannot separate two subsets each containing some SN vertices; (b) in a marked graph, there cannot exist two not connected G vertices having in their boundary sets exactly one SN vertex. From here, in some cases, we can identify which are the non-zero components of α.

The importance of identifying, for a given graph, which are the null elements of α and of Ω^{-1} lies in the possibility of using this information in the estimation stage. We have in mind the case where a marked conditional independence graph, associated with a certain applied problem, has been specified on the grounds of subject-matter considerations. For all pairs of vertices $\{i, j\}$ where an edge is missing, we know that $\Omega^{ij} = 0$ and at least one of α_i and α_j is zero. The use of this information in conjunction with the results established above can lead to a quite specific identification of the parameter structure; this process is exemplified in the next paragraph. The possibility of transferring the structure of the graph into constraints on the null elements of the parameter estimates can improve appreciably the estimation problem, avoiding the scan of a large set of compatible parameter patterns, and reducing variability of the estimates.

For an illustration, consider the graph in Figure 5.5, where the nature of vertex 5 is not yet specified. If we set $5 \in V_G$, from Corollary 5.21 we conclude that the graph is not admissible since the G nodes 2 and 5 would have on their boundary a single vertex belonging to V_{SN}, but they are not connected to each other. If we set $5 \in V_{SN}$, the graph becomes admissible. Since

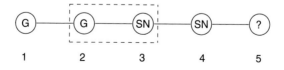

Figure 5.5 An example of a marked graph, where the labels G and SN denote Gaussian and extended skew-normal nodes, respectively, and the nature of the node marked '?' is discussed in the text. The dashed box indicates the nodes with possibly non-null α's in the joint 5-dimensional distribution.

$2 \in V_G$ separates $1 \in V_G$ from $\{3, 4, 5\} \subseteq V_{SN}$, condition (a) of Proposition 5.18 holds. The fact $Y_1 \perp\!\!\!\perp Y_{\{3,4,5\}}|Y_2$ is compatible both with $\alpha_1 = 0$ and with $\alpha_{\{3,4,5\}} = (0, 0, 0)^\top$. Corollary 5.19 indicates that we must have $\alpha_1 = 0$, and Proposition 5.20 implies that $\alpha_2 \neq 0$. Finally, the facts $Y_2 \perp\!\!\!\perp Y_5|Y_{\{1,3,4\}}$ and $Y_2 \perp\!\!\!\perp Y_4|Y_{\{1,3,5\}}$ lead us to say that the non-zero components of α can only be α_2 and α_3. Of these, we have established that α_2 is non-zero, while α_3 may be 0 or not.

5.3.6 Bibliographical notes

The multivariate ESN distribution has been studied by Adcock and Shutes (1999), Arnold and Beaver (2000a, Section 4) and Capitanio *et al.* (2003, Section 2 and Appendix). The first of these was motivated by application problems to quantitative finance, the main facts of which we shall recall in the next subsection. The other two papers present expressions for basic properties, such as the marginal and the conditional distributions, the moment generating function and lower-order moments, with inessential differences in the parameterization. Capitanio *et al.* (2003) also give expressions for the distribution function, the general expression of the cumulants and Mardia's coefficients.

The main target of Capitanio *et al.* (2003) is the development of a formulation for graphical models, of which § 5.3.5 represents an excerpt. Among the aspects not summarized here, this paper provides results for a parameter-based factorization of the likelihood function, which can simplify substantially complex estimation problems. Work on related graphical models has been presented by Stanghellini and Wermuth (2005). Capitanio and Pacillo (2008) propose a Wald-type test for the inclusion/exclusion of a single edge, and Pacillo (2012) explores the issue further.

5.3.7 Some applications

In quantitative finance, much work is developed under the assumption of multivariate normality, for convenience reasons. While it is generally agreed that normality is unrealistic, use of alternatives is often hampered by the lack of mathematical tractability. Adcock and Shutes (1999) have shown that various operations can be transferred quite naturally from the classical context of multivariate normal distribution to the ESN. They work with a parameterization which essentially is as in (5.30), with the introduction of an additional parameter, which leads to the ESN distribution. They obtain the moment generating function, lower-order moments and other basic properties. These results are employed to reconsider some classical optimality problems in finance within this broader context. As a specific instance, denote by R a vector of d asset returns, and examine the problem of optimal allocation of weights w among these assets, under the expected utility function

$$\psi(w) = 1 - \mathbb{E}\{\exp(-w^\top R/\theta)\},$$

where $\theta > 0$ is a parameter which expresses the risk appetite of the investor. If we assume that R has joint ESN distribution, then maximization of $\psi(w)$ corresponds to minimization of the moment generating function of type (5.62) evaluated at $t = -w/\theta$, more conveniently so after logarithmic transformation. The problem allows a simple treatment even in the presence of linear inequality constraints. The authors also deal with analogues of efficient frontier and market model. See also Adcock (2004) for closely related work and some empirical illustrations.

Carmichael and Coën (2013) formulate a model for asset pricing where the log-returns are jointly multivariate skew-normal and the stochastic discount factor is a polynomial transform of a reference component of them. The ensuring construction is sufficiently tractable for the authors to obtain analytic expressions for various quantities of interest and this 'sheds a new light on financial puzzles as the equity premium puzzle, the riskfree rate puzzle and could also be promising to deal with other well known financial anomalies' (Section 4).

Similarly to finance, in various other application areas the assumption of multivariate normality is often made for convenience and the SN or ESN distribution can be adopted as a more realistic and still tractable model. A case in point is represented by the work of Vernic (2006) in the context of insurance problems. For the evaluation of risk exposure, she considers the 'conditional tail expectation' (TCE), regarded as preferable to the more common indicator represented by value at risk. The TCE is defined for a random variable X as

$$\mathrm{TCE}_X(x_q) = \mathbb{E}\{X|X > x_q\}, \qquad x_q \in \mathbb{R},$$

which is much the same concept of mean residual life used in other areas.

For an ESN variable $Z \sim \mathrm{SN}(0, 1, \alpha, \tau)$, the TCE function can easily be computed via integration by parts lending, in the notation of §2.2,

$$
\begin{aligned}
\mathrm{TCE}_Z(z_q) &= \frac{1}{1 - \Phi(z_q; \alpha, \tau)} \int_{z_q}^{\infty} z\, \varphi(z; \alpha, \tau)\, \mathrm{d}z \\
&= \frac{1}{1 - \Phi(z_q; \alpha, \tau)} \left[\varphi(z_q; \alpha, \tau) + \delta\zeta_1(\tau)\, \Phi\left(-\sqrt{1 + \alpha^2}(z_q + \delta\tau)\right) \right]
\end{aligned}
$$

and the more general case $Y \sim \text{SN}(\xi, \omega^2, \alpha, \tau)$ is handled by the simple connection $\text{TCE}_Y(y_q) = \xi + \omega\,\text{TCE}_Z(z_q)$, where $z_q = \omega^{-1}(y_q - \xi)$.

For the purpose of optimal capital allocation in the presence of random losses $Y = (Y_1, \ldots, Y_d)^\top$, it is of interest to compute $\mathbb{E}\{Y_i | S > s_q\}$ where S is the total loss $S = \sum_i Y_i$ or more generally a linear combination of type $S = w^\top Y$. Under a multivariate ESN assumption for Y, Vernic (2006) shows how this computation can be performed in explicit form, leading to the TCE formula for capital allocation. The author also considers another allocation formula with respect to an alternative optimality criterion.

The multivariate SN distribution has been employed in a range of other application areas. Early usage of the bivariate SN distribution for data fitting includes the works of Chu *et al.* (2001) as a model for random effects in the analysis of some pharmacokinetics data and of Van Oost *et al.* (2003) in a study of soil redistribution by tillage. Many more applications have followed, however, often quite elaborate. Since they generally feature other modelling aspects or they intersect with the use of related distributions, these other contributions will be recalled at various places later on, many of them in §8.2 but also elsewhere.

5.4 Complements

Complement 5.1 (Canonical form and scatter matrices)   The construction of the canonical form $Z^* = H^\top(Y - \xi)$ of $Y \sim \text{SN}_d(\xi, \Omega, \alpha)$ in Proposition 5.13 involves implicitly the simultaneous diagonalization of Ω and $\Sigma = \text{var}\{Y\}$ to obtain matrix H. To see this, consider the equations

$$\Sigma h_j = \rho_j \Omega h_j, \qquad j = 1, \ldots, d, \qquad (5.78)$$

where $h_j \in \mathbb{R}^d$ and $\rho_j \in \mathbb{R}$. The solution of the jth equation is obtained when ρ_j and h_j are an eigenvalue and the corresponding eigenvector of $\Omega^{-1}\Sigma$. Since this matrix is similar to matrix M appearing in Proposition 5.13, it easily follows that h_j constitutes the jth column of H.

This reading of the canonical form establishes a bridge with the results of Tyler *et al.* (2009) based on the simultaneous diagonalization of two scatter matrices. Recall that, given a d-dimensional random variable X, a matrix-valued functional $V(X)$ is a scatter matrix if it is positive definite, symmetric and satisfies the property $V(b + A^\top X) = A^\top V(X) A$ for any vector $b \in \mathbb{R}^d$ and any non-singular $d \times d$ matrix A. The authors show how, from the diagonalization of two scatter matrices, information about the properties of a model can be established, as the vectors h_j identify important directions for inspecting data and they form an invariant coordinate system which, in the authors' words, 'can be viewed as a projection pursuit without the pursuit effort'. Also the ρ_j's provide information about the model; for instance, for an elliptical distribution they are all equal to each other. In our case, Ω and Σ represent two such scatter matrices.

Complement 5.2 (Regions of given probability)   For a skew-normal variable Z with specified parameter values, we examine the problem of finding the region $R_{\mathrm{SN}} \subset \mathbb{R}^d$ of smallest geometrical size such that $\mathbb{P}\{Z \in R_{\mathrm{SN}}\} = p$, for any given value $p \in (0, 1)$. First of all, notice that the problem is location and scale equivariant, so that it can be reduced to the case $Z \sim \mathrm{SN}_d(0, \bar{\Omega}, \alpha)$ where $\bar{\Omega}$ is a correlation matrix. Secondly, it is immediate to state that the solution must be of type

$$R_{\mathrm{SN}} = \{x : \varphi_d(x; \bar{\Omega}, \alpha) \geq f_0\},$$

where f_0 is a suitable value which ensures that $\mathbb{P}\{Z \in R_{\mathrm{SN}}\} = p$. The question is how to find f_0. Log-concavity of the SN density implies that R_{SN} is a convex set.

The analogous problem for a normal variable $X \sim \mathrm{N}_d(0, \Sigma)$ has a neat solution represented by the ellipsoid

$$\begin{aligned} R_{\mathrm{N}} &= \{x : x^{\top}\Sigma^{-1}x \leq c_p\} \\ &= \{x : 2\log \varphi_d(x; \Sigma) \geq -c_p - d\log 2\pi - d\log \det(\Sigma)\}, \end{aligned}$$

where c_p is the pth quantile of χ_d^2, on recalling that $X^{\top}\Sigma^{-1}X \sim \chi_d^2$. The region R_{N}, with Σ replaced by $\bar{\Omega}$, provides a region of probability p also for Z, since the χ_d^2 distribution is preserved, but in the SN case it does not represent the region of minimum geometrical size.

An exact expression of f_0 does not seem feasible, and an approximation must be considered. What follows summarizes the proposal of Azzalini (2001). As a first formulation, rewrite R_{N} replacing the normal density with $\varphi_d(x; \bar{\Omega}, \alpha)$, that is, consider the set

$$\tilde{R}_{\mathrm{SN}} = \{x : 2\log \varphi_d(x; \bar{\Omega}, \alpha) \geq -c_p - d\log(2\pi) - \log \det(\bar{\Omega})\} \qquad (5.79)$$

and let $\tilde{p} = \mathbb{P}\{Z \in \tilde{R}_{\mathrm{SN}}\}$.

To ease exposition, in the following we focus on the case $d = 2$, so that in (5.79) we have $c_p = -2\log(1 - p)$ and $\det(\bar{\Omega}) = 1 - \bar{\omega}_{12}^2$. Evaluation of $\tilde{p}$ can be performed via simulation methods, for any given choice of the parameter set. For a range of values from $p = 0.01$ to $p = 0.99$, say, the corresponding values $\tilde{p}$ can be estimated by the relative frequencies of $\tilde{R}_{\mathrm{SN}}$

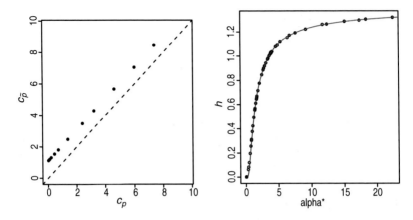

Figure 5.6 Construction of regions with given probability of SN
distribution. Left plot: $(c_p, c_{\tilde{p}})$ for a set of p values when the
distribution is bivariate skew-normal with parameters given in the
text. Right plot: points (α_*, h) for a set of parameter combinations
and interpolating curve.

in a set of sampled values. The left plot of Figure 5.6 refers to a simulation
of 10^6 values sampled with $\bar{\omega}_{12} = -0.5$ and $\alpha = (2, 6)^{\top}$; after converting
the $\tilde{p}$'s to the corresponding χ^2_2 quantiles, $c_{\tilde{p}}$'s say, these have been plotted
versus c_p. While the plotted points do not lie on the ideal identity line, they
are almost perfectly aligned along a line essentially parallel to the identity
line, with only a very slight upturn when c_p is close to 0. Hence with good
approximation we can write $c_{\tilde{p}} = c_p + h$ for some fixed h.

The pattern described above for that specific parameter set has been observed almost identically in a range of other cases, with different parameters. Invariably, the plotted points were aligned along the line $c_{\tilde{p}} = c_p + h$,
where h varied with $\bar{\Omega}$ and α. Another interesting empirical indication is
that h depends on the parameters only via α_* defined by (5.37). This is visible in the right panel of Figure 5.6, which plots a set of values of h, for
several parameter combinations, versus α_*; the interpolating line will be
described below. Therefore, a revised version of the approximate set is

$$\hat{R}_{\mathrm{SN}} = \{x : 2 \log \varphi_d(x; \bar{\Omega}, \alpha) \geq -c_p + h - d \log(2\pi) - \log \det(\bar{\Omega})\} \quad (5.80)$$

where h is a suitable function of α_*.

Some more numerical work shows that a good approximation to h is
provided by $h = 2 \log\{1 + \exp(-k_2/\alpha_*)\}$ where $k_2 = 1.544$, and this corresponds to the solid line in the right plot of Figure 5.6. This curve

visibly interpolates the points satisfactorily, with only a little discrepancy near the origin.

As a check of the validity of the revised formulation, we evaluate $\hat{p} = \mathbb{P}\{Z \in \hat{R}_{SN}\}$ with the same method described for $\tilde{p}$. The numerical outcome for the earlier case with $\bar{\omega}_{12} = -0.5$ and $\alpha = (2, 6)^\top$ is summarized in the following table:

p	0.01	0.05	0.3	0.5	0.8	0.95	0.99
$\hat{p}$	0.043	0.077	0.306	0.500	0.797	0.949	0.990

There is a satisfactory agreement between p and $\hat{p}$ for moderate and large p, which are the cases of main practical interest. A similar agreement between p and $\hat{p}$ has been observed with other parameter combinations.

The contour lines in Figures 5.2 and 5.3 have been chosen using this method, followed by suitable location and scale transformations. Hence, for instance, the region delimited by the curve labelled $p = 0.9$ has this probability, up to the described approximation, and has minimal area among the regions with probability 0.9.

The same procedure works also for other values of d, provided k_2 in the above expression of h is replaced by k_d, where $k_1 = 1.854$, $k_3 = 1.498$, $k_4 = 1.396$.

Complement 5.3 (Extension of Stein's lemma)   If $X \sim N(\mu, \sigma^2)$ and h is a differentiable function such that $\mathbb{E}\{|h'(X)|\} < \infty$, Stein's lemma states that

$$\mathrm{cov}\{X, h(X)\} = \sigma^2 \, \mathbb{E}\{h'(X)\} \ . \tag{5.81}$$

An extension to multivariate normal variables exists.

Adcock (2007) presents an extension of this result to the case of a multivariate ESN variable, which he developed as a tool for tackling an optimization problem in finance. His result was formulated in a parameterization essentially like (5.30) but, for homogeneity with the rest of our exposition, we recast the result for the parameterization $SN_d(\xi, \Omega, \alpha, \tau)$. Another difference is that the proof below makes use of the canonical form, which simplifies the logic of the argument. The transformation $Y^* = H^\top(Y - \xi)$ has been defined in Proposition 5.13 in connection with an SN distribution, hence with $\tau = 0$, but the same transformation works here, leading to $Y^* \sim SN_d(0, I_d, \alpha_{Z^*}, \tau)$; see Problem 5.14.

Lemma 5.22 *Let $Y \sim SN_d(\xi, \Omega, \alpha, \tau)$ and denote by $h(x)$ a real-valued function on $\mathbb{R}^d$ such that $h'_i(x) = \partial h(x)/\partial x_i$ is continuous and $\mathbb{E}\{|h'_i(Y)|\}$ is finite, for $i = 1, \ldots, d$. Then*

$$cov\{Y, h(Y)\} = \Omega \, \mathbb{E}\{\nabla h(Y)\} + (\mathbb{E}\{Y\} - \xi)\left(\mathbb{E}\{h(W)\} - \mathbb{E}\{h(Y)\}\right), \quad (5.82)$$

where $\nabla h(Y) = \left(h'_1(Y), \ldots, h'_d(Y)\right)^\top$, $W \sim N_d(\xi - \tau \omega \delta, \Omega - \omega \delta \delta^\top \omega)$, $\mathbb{E}\{Y\} = \xi + \zeta_1(\tau) \, \omega \, \delta$ as in (5.71); here as usual δ is given by (5.11) and ζ_1 by (2.20).

Proof Consider the canonical form $Z^* = H^\top(Y - \xi) \sim SN_d(0, I_d, \alpha_{Z^*}, \tau)$ where H is defined in Proposition 5.13. The variables $Z^*_1, \ldots, Z^*_d$ are mutually independent and the last $d-1$ components have $N(0, 1)$ distribution. Therefore, by the original Stein's lemma (5.81), we have

$$cov\{Z^*_i, h(Z^*)\} = \mathbb{E}\{h'_i(Z^*)\} \qquad (i = 2, \ldots, d),$$

first arguing conditionally on the remaining components and then, by independence, unconditionally. For $Z^*_1 \sim SN(0, 1, \alpha_*, \tau)$, we have

$$\mathbb{E}\{Z^*_1 \, h(Z^*)\} = \int_{\mathbb{R}^{d-1}} \prod_{j=2}^d \varphi(z_i) \left[\int_{\mathbb{R}} h(z) \, z_1 \, \varphi(z_1; \alpha_*, \tau) \, dz_1 \right] dz_2 \cdots dz_d$$

where $\varphi(z_1; \alpha_*, \tau)$ is given by (2.39). Expansion of the inner integral by parts lends

$$\int_{\mathbb{R}} \frac{\partial}{\partial z_1} h(z) \varphi(z_1; \alpha_*, \tau) \, dz_1 + \frac{\alpha_* \varphi(\tau)}{\Phi(\tau)} \int_{\mathbb{R}} h(z) \, \varphi\left((z_1 + \tau \delta(\alpha_*)) \sqrt{1 + \alpha_*^2}\right) dz_1,$$

where $\delta(\cdot)$ is given by (2.6). When this expression is inserted back in the d-dimensional integral, we get

$$\mathbb{E}\{Z^*_1 h(Z^*)\} = \mathbb{E}\{h'_1(Z^*)\} + \delta(\alpha_*) \, \zeta_1(\tau) \int_{\mathbb{R}^d} h(u) \, \varphi_d(u - \mu_U; \Omega_U) \, du,$$

where $\mu_U = (-\tau \delta(\alpha_*), 0, \ldots, 0)^\top$ and $\Omega_U = \text{diag}(1 - \delta(\alpha_*)^2, 1, \ldots, 1)$. On recalling that $\delta(\alpha_*) = \delta_*$ where δ_* is defined by (5.38), write

$$\mathbb{E}\{Z^*_1 \, h(Z^*)\} = \mathbb{E}\{h'_1(Z^*)\} + \zeta_1(\tau) \delta_* \mathbb{E}\{h(U)\},$$

where $U \sim N_d(\mu_U, \Omega_U)$, and from (5.71) we obtain

$$cov\{Z^*, h(Z^*)\} = \mathbb{E}\{\nabla h(Z^*)\} + \mathbb{E}\{Z^*\}\left(\mathbb{E}\{h(U)\} - \mathbb{E}\{h(Z^\top)\}\right).$$

Since $Z^* = H^\top(Y - \xi)$, then $H^\top \Omega H = I_d$, and so also $(H^\top)^{-1} H^{-1} = \Omega$. Moreover, since $\Omega_U = I_d - \zeta_1(\tau)^{-2} \mathbb{E}\{Z^*\} \mathbb{E}\{Z^*\}^\top$, we obtain $(H^\top)^{-1} \Omega_U H^{-1} = \Omega - \omega \delta \delta^\top \omega$, bearing in mind (5.71). From these facts and

$$\operatorname{cov}\{Y, h(Y)\} = \operatorname{cov}\{Y - \xi, h(Y)\} = (H^\top)^{-1} \operatorname{cov}\left\{Z^*, h(\xi + (H^\top)^{-1} Z^*)\right\},$$

we arrive at (5.82). QED

Problems

5.1 Prove Proposition 5.1.

5.2 Confirm that the distribution of $Z = (Z_1, \ldots, Z_d)$ whose components are defined by (5.19) is $\mathrm{SN}_d(0, \bar{\Omega}, \alpha)$, where $\bar{\Omega}$ and α are given by (5.21) and (5.22) (Azzalini and Dalla Valle, 1996).

5.3 Show that, for any choice of $\bar{\Omega}$ and α, there is a choice of $\bar{\Psi}$ and δ in (5.16)–(5.19) leading to the distribution $\mathrm{SN}_d(0, \bar{\Omega}, \alpha)$. From here show how the parameterization (ξ, Ψ, λ) of (5.30) can be mapped to (ξ, Ω, α) of (5.3), and conversely (Azzalini and Capitanio, 1999, Appendix of the full version).

5.4 In § 5.1.3 it is stated that $\bar{\Omega}$ and δ are not variation independent; hence not all choices $(\bar{\Omega}, \delta)$ are admissible. Show that a necessary and sufficient condition for their admissibility is $\bar{\Omega} - \delta \delta^\top > 0$.

5.5 Check (5.27). Also, show that in case $h = 1$, the expression reduces to

$$\alpha_{1(2)} = \left(1 + \alpha_{2*}^2 - u^2\right)^{-1/2} (\alpha_1 + u)$$

where $u = \bar{\Omega}_{12} \alpha_2$ and $\alpha_{2*}^2 = \alpha_2^\top \bar{\Omega}_{22} \alpha_2$. Finally, show that $\alpha_{1(2)} = \lambda_1$, the first component of vector λ in (5.20).

5.6 Confirm that the parameters of (5.40) are as given by (5.41)–(5.44).

5.7 Show that α_X in (5.43) can be written as

$$\left(1 + \alpha^\top \omega^{-1} \left(\Omega - \Omega A \Omega_X^{-1} A^\top \Omega\right) \omega^{-1} \alpha\right)^{-1/2} \omega_X \Omega_X^{-1} A^\top \Omega \omega^{-1} \alpha.$$

5.8 Confirm the statement at the end of § 5.1.2 that the sum of two independent multivariate SN variables, both with non-zero slant, is not of SN type.

5.9 Consider the variable (U, Z), where $U \in \mathbb{R}$ and $Z = (Z_1, \ldots, Z_d)^\top \in \mathbb{R}^d$, with joint density

$$f(u, z) = \frac{(1 + \alpha_*^2)^{1/2}}{(2\pi)^{(d+1)/2} \det(\bar{\Omega})^{1/2}} \exp\left\{-\tfrac{1}{2}\left(u^2(1 + \alpha_*^2)\right.\right.$$
$$\left.\left. -2(1 + \alpha_*^2)^{1/2} |u| \alpha^\top z + z^\top \bar{\Omega}^{-1} z + (\alpha^\top z)^2\right)\right\}$$

where $\bar{\Omega} > 0$ is a correlation matrix and $\alpha_*^2 = \alpha^\top \bar{\Omega} \alpha$. Show the following: (a) marginally, $U \sim N(0,1)$ and $Z \sim SN_d(0, \bar{\Omega}, \alpha)$; (b) Z and U are independent if and only if $\alpha = 0$; (c) $\text{cov}\{U, Z_i\} = 0$ for $i = 1, \ldots, d$.

5.10 Consider a bivariate SN distribution with location $\xi = 0$, $\text{vech}(\Omega) = (1, r, 1)^\top$ and $\alpha = a(-1, 1)^\top$ where $r \in (-1, 1)$ and $a \in \mathbb{R}$. Show that, if $a \to \infty$, then $\delta \to \frac{1}{2}\sqrt{1-r}(-1, 1)^\top$. If further $r \to 1$, then $\delta \to 0$ and correspondingly $\gamma_1 \to 0$ for each marginal. Examine the numerical values of γ_1 and the contour lines plot of the density in the case $r = 0.9$ and $a = 100$. Comment on the qualitative implications.

5.11 For a $(d+1)$-dimensional normal distribution as in (5.14), consider the conditional distribution of X_0 under two-sided constraint of X_1, that is $Z = (X_0 | a < X_1 < b)$ where a and b are arbitrary, provided $a < b$. Obtain the density function and the lower-order moments of Z, specifically the marginal coefficients of skewness and kurtosis (Kim, 2008).

5.12 Suppose that $X \sim SN_d(\xi, \Omega, \alpha)$ is partitioned as $X = (X_1^\top, X_2^\top)^\top$, where X_1 has dimension h. Then show, without using (5.67), that the distribution of X_2 conditionally on $X_1 = x_1$ is $SN_{d-h}(\xi_{2\cdot 1}, \Omega_{22.1}, \alpha_{2\cdot 1}, \tau_c)$, of which the first three parameter components are as in (5.67) and $\tau_c = \alpha_{1(2)}^\top \omega_1^{-1}(x_1 - \xi_1)$, where $\alpha_{1(2)}$ is given by (5.27).

5.13 Show that the additive representation (5.70) of a multivariate ESN distribution is equivalent to representation (5.68).

5.14 Extend the idea of the canonical form of § 5.1.8 to the ESN case. Specifically, if $Y \sim SN_d(\xi, \Omega, \alpha, \tau)$, show that there exist a matrix H such that $Z^* = H^\top(Y - \xi) \sim SN_d(0, I_d, \alpha_{Z^*}, \tau)$, so that the distribution of Z^* can be factorized as the product of $d - 1$ standard normal densities and that of $SN(0, 1, \alpha_*, \tau)$, where α_{Z^*} and α_* are as for the SN case. Use this result to derive the final expressions in (5.74) and (5.75).

5.15 If $X = (X_1, \ldots, X_d)^\top \sim N_d(0, \Omega)$ where $\Omega > 0$, Šidák (1967) has shown that the inequality

$$\mathbb{P}\{|X_1| \le c_1, \ldots, |X_d| \le c_d\} \ge \prod_{i=1}^{d} \mathbb{P}\{|X_i| \le c_i\}$$

holds for any positive numbers $c_1, \ldots, c_d$. Prove that the same inequality holds when $X \sim SN_d(0, \Omega, \alpha)$, and that for any $p \in [0, 1]$ the choice of a sequence $c_1, \ldots, c_d$ such that $\mathbb{P}\{|X_1| \le c_1, \ldots, |X_d| \le c_d\} = 1 - p$ does not depend on the parameter α.

5.16 Show that the set of distributions $SN_5(\xi, \Omega, \alpha, \tau)$ compatible with the marked conditional independence graph depicted below must satisfy the condition $\{\alpha_1 = \alpha_5 = 0\} \cap \{\alpha_2 \ne 0\}$. Also, show that changing the graph to $1 \in V_{SN}$ would make it incompatible with the above ESN assumption.

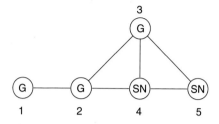

5.17 Consider the density

$$f(x) = 2\,\varphi_2(x; \bar{\Omega}, \alpha)\,\Phi\{\lambda\,(x_1^2 - x_2^2)\}, \qquad x = (x_1, x_2) \in \mathbb{R}^2,$$

where $\bar{\Omega}$ is a correlation matrix, $\alpha \in \mathbb{R}^2$ and $\lambda \in \mathbb{R}$, that is, a density similar to the one in (1.30) but with the bivariate normal density replaced by a bivariate SN. Confirm that $f(x)$ is a proper density and show that, if X has density $f(x)$ with $\alpha = a\,(1, 1)^\top$ for some $a \in \mathbb{R}$, then $X^\top \bar{\Omega}^{-1} X \sim \chi_2^2$.

6

Skew-elliptical distributions

with emphasis on the skew-t family

6.1 Skew-elliptical distributions: general aspects

At the beginning of Chapter 4 we argued that we need to consider a base density whose tails can be regulated by some parameter. The same motivation holds in the multivariate context as well, even if the notion of 'tail' must now be suitably adapted. In this chapter, we apply the modulating-symmetry process of Chapter 1 to symmetric multivariate distributions with the structure described next.

6.1.1 A summary of elliptically contoured distributions

The class of elliptically contoured distributions, or more briefly *elliptical distributions*, is connected to the idea that the density is constant on ellipsoids. The theory of this area is much developed, and we recall here only the main facts, under the restriction of continuous random variables, which is our case of interest. We refer the reader to some standard account, such as Fang and Zhang (1990) or Fang *et al.* (1990), for a detailed treatment and proofs.

For a positive integer d, consider a function $\tilde{p}$ from $\mathbb{R}^+$ to $\mathbb{R}^+$ such that

$$k_d = \int_0^\infty r^{d-1} \, \tilde{p}(r^2) \, \mathrm{d}r < \infty. \tag{6.1}$$

Then a d-dimensional continuous random variable X is said to have an elliptical distribution, with *density generator* $\tilde{p}$, if its density is of the form

$$p(x; \mu, \Sigma) = \frac{c_d}{\det(\Sigma)^{1/2}} \, \tilde{p}\{(x - \mu)^\top \Sigma^{-1}(x - \mu)\}, \qquad x \in \mathbb{R}^d, \tag{6.2}$$

where $\mu \in \mathbb{R}^d$ is a location parameter, Σ is a symmetric positive-definite $d \times d$ scale matrix and $c_d = \Gamma(d/2)/(2 \pi^{d/2} k_d)$. In this case we shall write $X \sim \mathrm{EC}_d(\mu, \Sigma, \tilde{p})$. If $\Sigma = I_d$, the distribution is said to be *spherical*.

Clearly, density (6.2) is constant over the set of points x such that

$$(x - \mu)^\top \Sigma^{-1}(x - \mu) = \text{constant},$$

which is the equation of an ellipsoid. Another obvious remark, but crucial for us, is that elliptical densities are centrally symmetric around their location parameter.

Proposition 6.1   *The following properties hold for $X \sim \text{EC}_d(\mu, \Sigma, \tilde{p})$.*

(a) *If A is a full-rank $d \times d$ matrix and c is a d-vector, then*

$$Y = c + A^\top X \sim \text{EC}_d(c + A^\top \mu, A^\top \Sigma A, \tilde{p}). \qquad (6.3)$$

(b) *There exists a random vector S, uniformly distributed on the unit sphere in $\mathbb{R}^d$, and a continuous positive random variable R, independent of S, having density function*

$$f_R(r) = k_d^{-1} \, r^{d-1} \, \tilde{p}(r^2), \qquad 0 < r < \infty,$$

such that

$$X \overset{\mathrm{d}}{=} \mu + RL^\top S \qquad (6.4)$$

where $L^\top L = \Sigma$; density f_R is called the radial distribution.

(c) *The previous statement implies that $(X - \mu)^\top \Sigma^{-1}(X - \mu) \overset{\mathrm{d}}{=} R^2$.*

(d) *If R has finite second moment, then*

$$\mathbb{E}\{X\} = \mu, \qquad var\{X\} = \mathbb{E}\{R^2/d\} \, \Sigma. \qquad (6.5)$$

(e) *If Σ is a diagonal matrix, the components of X are independent if and only if X has a multinormal distribution.*

(f) *If X_1 is a random vector obtained by selecting h components of X, for some $0 < h < d$, and μ_1 and Σ_{11} are the blocks of μ and Σ corresponding to the selected indices, then*

$$X_1 \sim \text{EC}_h(\mu_1, \Sigma_{11}, \tilde{p}_h) \qquad (6.6)$$

where the density generator $\tilde{p}_h$ depends on $\tilde{p}$ and h only, not on the specific choice of the components of X.

(g) *If X_2 is the vector obtained from X by removing the components in X_1, then the conditional distribution of X_2 given that $X_1 = x_1$ is, in an obvious notation for the partitions of μ and Σ,*

$$\text{EC}_{d-h}(\mu_2 + \Sigma_{21}\Sigma_{11}^{-1}(x_1 - \mu_1), \, \Sigma_{22 \cdot 1}, \, \tilde{p}_{Q(x_1)}) \qquad (6.7)$$

where $\Sigma_{22 \cdot 1} = \Sigma_{22} - \Sigma_{21}\Sigma_{11}^{-1}\Sigma_{12}$ and the density generator $\tilde{p}_{Q(x_1)}$ depends on x_1 only through $Q(x_1) = (x_1 - \mu_1)^\top \Sigma_{11}^{-1}(x_1 - \mu_1)$.

(h) The density generator $\tilde{p}_{Q(x_1)}$ in (6.7) does not depend on x_1 if and only if X is multinormal.

From representation (6.4), we can think that a value sampled from X is generated as follows: a uniformly random direction is drawn from S, and a point along this direction is projected to distance R; this is followed by linear transformation L and a final shift μ. This interpretation illustrates how all parametric families which comprise the elliptical class with given dimension d essentially differ for the effect of R only, which causes the spacings of the contour lines of the density to differ from one family to another, but they are otherwise the same.

The expression of the conditional distribution (6.7) resembles closely the analogous one for the multinormal case, but in general the conditional variance, to be computed from (6.5), depends on the radial distribution which for the conditional distribution depends on $\tilde{p}_{Q(x_1)}$, hence on x_1.

The class of elliptical distributions can be thought of as the union of a set of parametric families, one for any given density generator. The most prominent such family is obtained by choosing $\tilde{p}(u) = \exp(-u/2)$ so that $c_d = (2\pi)^{-d/2}$, leading to the multivariate normal density. Many other important multivariate distributions belong to the elliptical class. Among these, mention is due of the multivariate Pearson type VII distribution whose density generator and normalizing constant are

$$\tilde{p}(u) = (1 + u/v)^{-M}, \qquad c_d = \frac{\Gamma(M)}{(\pi v)^{d/2}\,\Gamma(M - d/2)}, \qquad (6.8)$$

respectively, provided $v > 0$ and $M > d/2$. The special importance of this family lies in that, when $M = (d + v)/2$, it leads to the commonly adopted form of multivariate Student's t density, that is, if $\mu = 0$,

$$t_d(x; \Sigma, v) = \frac{\Gamma((v + d)/2)}{(v\pi)^{d/2}\,\Gamma(v/2)\,\det(\Sigma)^{1/2}} \left(1 + \frac{x^\top \Sigma^{-1} x}{v}\right)^{-\frac{v+d}{2}}, \qquad x \in \mathbb{R}^d. \quad (6.9)$$

An important subset of the elliptical class is represented by the scale mixtures of normals, that is, those which can be generated as

$$X = \mu + W/\sqrt{V}, \qquad (6.10)$$

where $W \sim N_d(0, \Sigma)$ and V is a univariate positive random variable, independent of W. It is easy to see that a variable X so constructed has elliptical distribution. Among the many families of this type, the more familiar case is represented by the Student's t (6.9) which is obtained when $V \sim \chi_v^2/v$. Another example is the slash distribution obtained when $\sqrt{V} = U^{1/q}$, where $U \sim U(0, 1)$ and q is a positive parameter which regulates the tail weight.

An important property of scale mixtures of normals is the so-called consistency under marginalization, which means that the density generator after marginalization remains unchanged, except that d must be adjusted accordingly when it appears explicitly. This consistency property holds, among others, for the multivariate Student's t, given its stochastic representation just recalled. A similar fact does not hold, instead, for the elliptical class generated from Subbotin's distribution (4.1), obtained by replacing $|x|^\nu$ with $(x^\top \Sigma^{-1} x)^{\nu/2}$. The marginal components of this multivariate Subbotin distribution are still elliptical, because of (6.6), but they are not of Subbotin type themselves (Kano, 1994).

6.1.2 Skew-elliptical distributions: basic facts

Starting from the base density (6.2), the range of distributions which can be obtained by the general form of the modulation mechanism (1.2) is enormous. For the development of manageable parametric families, we need to narrow down our investigation to some more structured forms. One such construction is the direct extension of (1.25) to the present context:

$$f(x) = 2\, p_0(x)\, G_0(\alpha^\top x), \qquad x \in \mathbb{R}^d, \tag{6.11}$$

where now p_0 is an elliptical density with location at the origin, $\alpha \in \mathbb{R}^d$ is a vector of arbitrary constants and G_0 is as in (1.2). Although this is a legitimate family of distributions, with a simple formulation, we turn our attention to a somewhat different construction, for reasons which will emerge in the development.

Recall the stochastic representation of an SN_d variate via the conditioning mechanism $(X_0|X_1 > 0)$ applied to a $(d+1)$-dimensional normal variate $(X_0^\top, X_1)^\top$, and apply the same process to an elliptical variable with generator $\tilde{p}$. Specifically, start by introducing, similarly to (5.14),

$$X = \begin{pmatrix} X_0 \\ X_1 \end{pmatrix} \sim \mathrm{EC}_{d+1}\left(0, \Omega^*, \tilde{p}\right), \qquad \Omega^* = \begin{pmatrix} \bar{\Omega} & \delta \\ \delta^\top & 1 \end{pmatrix}, \tag{6.12}$$

where Ω^* is a full-rank correlation matrix, and consider the distribution of $Z \stackrel{\mathrm{d}}{=} (X_0|X_1 > 0)$. The density function of Z can be obtained by proceeding as in (1.28), for the conditioning set $(0, \infty)$. The marginal density of X_0 is still of elliptical type because of (6.6), denoted p_0, and

$$\mathbb{P}\{X_1 > 0 | X_0 = x\} = \int_0^\infty p_{Q(x)}(u; \delta^\top \bar{\Omega}^{-1} x, 1 - \delta^\top \bar{\Omega}^{-1} \delta)\, \mathrm{d}u$$

where we have used (6.7) and $p_{Q(x)}$ denotes the density generated by $\tilde{p}_{Q(x)}$; here $Q(x) = x^\top \bar{\Omega}^{-1} x$. Finally, taking into account central symmetry of elliptical densities, we obtain

$$f_z(x) = 2\, p_0(x)\, \mathbb{P}\{X_1 > 0 | X_0 = x\}$$
$$= 2\, p_0(x)\, P_{Q(x)}(\alpha^\top x), \tag{6.13}$$

where α is defined as in (5.12) and $P_{Q(x)}$ is the distribution function of $p_{Q(x)}$.

In general, $P_{Q(x)}(\alpha^\top x)$ is a non-linear function of x. Linearity occurs only in the special case when (6.12) is of normal type, since then $P_{Q(x)}(\cdot)$ does not depend on x, by Proposition 6.1(h). In this case we return to the SN distribution.

At first glance, (6.13) does not look like an instance of (1.2), but this is indeed the case. Since $P_{Q(x)}(\cdot)$ is symmetric about 0 and $Q(-x) = Q(x)$, it follows that $G(x) = P_{Q(x)}(\alpha^\top x)$ satisfies (1.4). Therefore, (6.13) can be written in the form (1.3) and, by Proposition 1.2, also as (1.2).

An implication is that Proposition 1.3 applies here, with $G(x)$ given by the conditional probability term in (6.13). Therefore, if X is as in (6.12), then both variables

$$Z' = (X_0 | X_1 > 0), \qquad Z = \begin{cases} X_0 & \text{if } X_1 > 0, \\ -X_0 & \text{if } X_1 \leq 0 \end{cases} \tag{6.14}$$

have distribution (6.13), which establishes a stochastic representation via a conditioning mechanism. Also, from (6.4) and the modulation invariance property (1.12), we have that $Z^\top \bar{\Omega}^{-1} Z \overset{d}{=} R^2$.

To introduce location and scale parameters, consider the transformation $Y = \xi + \omega Z$ where ξ and ω are as in (5.2) and Z has distribution (6.13). The density function of Y at $x \in \mathbb{R}^d$ is

$$f_Y(x) = \frac{2}{\det(\omega)}\, p_0(z)\, P_{Q(z)}(\alpha^\top z), \qquad z = \omega^{-1}(x - \xi), \tag{6.15}$$

where $Q(z) = z^\top \bar{\Omega}^{-1} z$. We shall say that Y has a *skew-elliptical* distribution and write $Y \sim \mathrm{SEC}_d(\xi, \Omega, \tilde{p})$.

The graphical appearance of the skew-elliptical density resembles that of the skew-normal distribution, illustrated in Figures 5.1, 5.2 and 5.3 for $d = 2$, but the contour lines are spread differently, due to the effect of changing the radial distribution. Specific illustrations will be given for the multivariate ST distribution in §6.2.

The construction of the skew-elliptical distributions (6.13) has been built starting from the so-called representation by conditioning of the SN distribution and replacing the normal distribution with the elliptical one. A similar connection exists for the additive representation (5.19): it can be proved that, if the assumption of a normal distribution for (U_0, U_1) in (5.16) is replaced by the assumption of an elliptical distribution, we arrive again at density (6.13); see Problem 6.1. Note that in the non-Gaussian case U_0 and U_1 are uncorrelated but not independent. Furthermore, the third type of stochastic representation, via minima and maxima, also holds for SEC variables, analogously to (5.25); see Problem 6.2.

6.1.3 Scale mixtures of SN variables

Consider now a $(d + 1)$-dimensional normal variable W, partitioned into W_0 and W_1 of dimension d and 1, respectively. Define X in the form (6.10) with $\mu = 0$, and partition X similarly to W. The conditional distribution of X_0 given that $X_1 > 0$ is of skew-elliptical type, since the normalization to 1 of the scale parameter of X_1 is irrelevant in this process. This amounts to consider the distribution of $W_0 / \sqrt{V}$ given $W_1 > 0$. Given the independence of V, it is equivalent to consider the distribution of $Z \stackrel{\text{d}}{=} (W_0|W_1 > 0)$, which is skew-normal, followed by division by $\sqrt{V}$.

To conclude, if the parent elliptical distribution of the skew-elliptical density (6.13) belongs to the scale mixtures of normals, then this density up to a vector of scale factors can be obtained as a scale mixture of skew-normal variables, with the same mixing variable V. Incorporating location and scale parameters, we arrive at considering variables of the type

$$Y = \xi + V^{-1/2}Z = \xi + S\,Z, \tag{6.16}$$

where $Z \sim SN_d(0, \Omega, \alpha)$ and $S = V^{-1/2}$ is usually chosen so that its median is near to 1. The density function of Y is obtained by integrating its conditional distribution given that $V = v$, that is

$$f_Y(x) = \int_0^\infty \varphi_d(x - \xi; v^{-1}\Omega, \alpha)\, f_V(v)\, \mathrm{d}v \tag{6.17}$$

if f_V denotes the density function of V. In some favourable cases, the integration in (6.17) can be carried out in an explicit form.

Representation (6.16) allows us to derive in a simple way a set of results. The first one is that, provided $\mathbb{E}\{S\}$ exists,

$$\mathbb{E}\{Y\} = \xi + \mathbb{E}\{S\}\, b\,\omega\,\delta$$

where b, ω and δ are as in Chapter 5. For higher moments, write $\mathbb{E}\{X^{(m)}\}$

to denote any moment of order m of a variable X. Then, provided $\mathbb{E}\{S^m\}$ exists, we have

$$\mathbb{E}\{(Y - \xi)^{(m)}\} = \mathbb{E}\{S^m\}\, \mathbb{E}\{Z^{(m)}\} \tag{6.18}$$

which says that the inflating factor $\mathbb{E}\{S^m\}$ depends on m only, not on the specific choice of indices and exponents. For the variance we obtain

$$\text{var}\{Y\} = \mathbb{E}\{S^2\}\Omega - \mathbb{E}\{S\}^2\, b^2 \omega \delta \delta^\top \omega.$$

Another implication of (6.16) is that the class of scale mixtures of skew-normal variables is closed with respect to marginalization and to affine transformations, since these properties hold for the SN family.

The canonical form of Y corresponds in essence to the transformed variable

$$Y^* = \xi + V^{-1/2}Z^* = \xi + S\, Z^*, \tag{6.19}$$

where V and S are as in (6.16) and $Z^* \sim \text{SN}_d(0, I_d, \alpha_{Z^*})$ is the canonical form of Z. The properties of Y^* are largely those of Z^*, discussed in § 5.1.8, with the predictable difference that its components are not independent, but only uncorrelated; the last fact can be checked using (6.18). It is immediate to verify that (6.19) can be obtained as $Y^* = H^\top(Y - \xi)$, where H is defined in Proposition 5.13.

The canonical form provides a route to compute Mardia's measures of multivariate skewness and excess kurtosis of Y, since they are invariant with respect to affine transformations. Starting from (1.2) and (2.9) of Mardia (1974) and using (6.18), one arrives after lengthy algebra at

$$\gamma_{1,d}^M = (\gamma_1^*)^2 + \frac{3(d-1)}{\sigma_*^2\, \psi_2^2}\, (\psi_3 - \psi_1\, \psi_2)^2\, \frac{2}{\pi}\delta_*^2, \tag{6.20}$$

$$\begin{aligned}\gamma_{2,d}^M = \beta_2^* &+ (d-1)(d+1)\, \psi_2^{-2}\, \psi_4 \\ &+ \frac{2(d-1)}{\sigma_*^2\, \psi_2}\left\{\psi_4 + \left(\psi_1^2\, \psi_2 - 2\psi_1\, \psi_3\right)\frac{2}{\pi}\delta_*^2\right\} - d(d+2),\end{aligned} \tag{6.21}$$

provided

$$\psi_m = \mathbb{E}\{S^m\} = \mathbb{E}\{V^{-m/2}\}, \qquad m = 1,\ldots,4$$

exist; here the quantities γ_1^*, β_2^* and σ_*^2 refer, in an obvious notation, to the component Y_1^* of Y^* and δ_* is defined in (5.38).

6.1.4 Bibliographic notes

An initial formulation of skew-elliptical distributions of the linear form (6.11) has been discussed briefly by Azzalini and Capitanio (1999, Section 7). Moving from a 'slightly different' viewpoint, Branco and Dey (2001) have put forward a construction based on the conditioning argument which leads to distributions (6.13). Additional work in this direction has been developed by Azzalini and Capitanio (2003). One of the questions tackled in this paper is the connection between the Branco–Dey formulation and the form (1.2) when the base density is elliptical, showing that at least in some important special cases distributions (6.13) are of type (1.2); these include the multivariate skew-*t* family to be discussed shortly. A result not discussed here is that a representation of type (6.4) exists with S not uniform on the sphere. One of the results of Azzalini and Regoli (2012a) confirms the conjecture that all distributions (6.13) are of type (1.2) with elliptical base density.

In the paper of Branco and Dey (2001) special emphasis is given to the subset of skew-elliptical densities generated as scale mixtures of multivariate skew-normal variables, and various cases are exemplified. This class has been examined further by Lachos *et al.* (2010a), under the label 'skew-normal/independent distributions', with special emphasis on computational aspects for model fitting. Kim and Genton (2011) obtain the characteristic function for this class and other related distributions. Capitanio (2012) has extended the notion of canonical form of skew-normal variates to their scale mixtures, leading to the above-quoted expressions of $\gamma_{1,d}^M$ and $\gamma_{2,d}^M$.

The work of Genton and Loperfido (2005) examines distributions with base density of elliptical type and modulation factor expressed in the form (1.4), and they develop a number of results such as modulation invariance for this form of skew-elliptical distributions. Further developments along this line have led to the paper by Wang *et al.* (2004) recalled in Chapter 1. There is substantial overlap of the last two cited papers with the work of Azzalini and Capitanio (2003), but they have been developed independently and at about the same time, in spite of the discrepancy in the publication dates.

Several other results, some very technical, on the distribution theory of skew-elliptical distributions have been obtained by B. Q. Fang in a series of papers (2003; 2005a; 2005b; 2006; 2008).

6.2 The multivariate skew-*t* distribution

6.2.1 *Some equivalent constructions*

In Chapter 4 we introduced the ST distribution via the ratio of an SN(0, 1, α) variate and an independent variable $\sqrt{V}$, where $V \sim \chi_\nu^2/\nu$. The natural multivariate extension of this construction is

$$Z = V^{-1/2} Z_0 \tag{6.22}$$

where now $Z_0 \sim \text{SN}_d(0, \bar{\Omega}, \alpha)$, independent of V, for some non-singular correlation matrix $\bar{\Omega}$ and $\alpha \in \mathbb{R}^d$. This is the classical genesis of the multivariate Student's t with density (6.9), when $Z_0 \sim \text{N}_d(0, \Sigma)$, for some $\Sigma > 0$.

The density of Z in (6.22) is obtained by integrating out the distribution of V similarly to (6.17). Using Corollary B.3 on p. 233 to express the integral, we obtain that the density at $z \in \mathbb{R}^d$ is

$$f_z(z) = 2\, t_d(z; \bar{\Omega}, \nu)\, T\left(\alpha^\top z \sqrt{\frac{\nu + d}{\nu + Q(z)}}; \nu + d\right), \tag{6.23}$$

where $Q(z) = z^\top \bar{\Omega}^{-1} z$ and $T(\cdot; \rho)$ denotes the univariate Student's t distribution function on ρ d.f. This is a direct extension of the process leading to (4.11) when $d = 1$.

Density (6.23) is of type (1.2) with a $t_d(z; \bar{\Omega}, \nu)$ base density and G_0 given by the $T(\cdot; \nu + d)$ distribution function evaluated at a non-linear odd function $w(z)$. As ν diverges, density (6.23) converges to the multivariate SN density (5.1).

Distribution (6.23) can be obtained also as an instance of the skew-elliptical distributions (6.13), when the distribution of X in (6.12) is $(d+1)$-dimensional Student's t; see Problem 6.3. A third way of arriving at (6.23) is via an additive representation, since we have seen that this exists in general for skew-elliptical distributions.

To introduce location and scale parameters, consider the transformation $Y = \xi + \omega Z$, similarly to (5.2). We shall say that Y has a multivariate *skew-t* (ST) distribution and write $Y \sim \text{ST}_d(\xi, \Omega, \alpha, \nu)$, where $\Omega = \omega \bar{\Omega} \omega$. The density function of Y at $x \in \mathbb{R}^d$ is

$$f_Y(x) = \det(\omega)^{-1} f_z(z), \qquad z = \omega^{-1}(x - \xi). \tag{6.24}$$

6.2.2 Main properties

Representation (6.22) allows us to derive very simply several important properties. Using (6.18), it is immediate that, for $Y = \xi + \omega Z$,

$$\mu = \mathbb{E}\{Y\} = \xi + \omega \mu_z, \qquad \text{if } v > 1, \qquad (6.25)$$

$$\Sigma = \text{var}\{Y\} = \frac{v}{v-2}\Omega - \omega \mu_z \mu_z^\top \omega, \qquad \text{if } v > 2, \qquad (6.26)$$

where $\mu_z = b_v \delta$ with b_v given by (4.15) and δ is given by (5.11).

For the distribution function of Z, we argue as for (5.69) combined with consideration of (6.22) and obtain

$$\mathbb{P}\{Z \le z\} = 2\,\mathbb{P}\left\{V^{-1/2}\begin{pmatrix} X_0 \\ -X_1 \end{pmatrix} \le \begin{pmatrix} z \\ 0 \end{pmatrix}\right\} = 2\,\mathbb{P}\left\{T' \le \begin{pmatrix} z \\ 0 \end{pmatrix}\right\}, \qquad (6.27)$$

where T' is a $(d+1)$-dimensional Student's t on v d.f. and scale matrix similar to Ω^* in (5.14) with δ replaced by $-\delta$.

Another application of (6.22) gives the distribution of an affine transformation of Y:

$$X = c + A^\top Y \sim \text{ST}_h(\xi_X, \Omega_X, \alpha_X, v), \qquad (6.28)$$

where c and A are as in (5.40) and the first three parameter components of X are as in (5.41)–(5.43). Similarly to the SN case in § 5.1.4, the marginal distribution of Y_1 constituted by the first h components of Y is

$$Y_1 \sim \text{ST}_h(\xi_1, \Omega_{11}, \alpha_{1(2)}, v), \qquad (6.29)$$

where the first three parameter components are as in (5.26) and (5.27). Note that the property of closure under affine transformation (6.28) occurs thanks to the construction (6.22) with a common denominator $V^{-1/2}$ for all components, which involves a single tail weight parameter v.

For a quadratic form of type $Q = Z^\top B Z$, where B is a symmetric $d \times d$ matrix, write $Q = Z_0^\top B Z_0 / V$. Corollary 5.9 ensures that, for appropriate choices of B, the distribution of $Z_0^\top B Z_0$ is χ_q^2, for some value q, and correspondingly Q/q is distributed as a Snedecor's $F(q, v)$. A case of special interest is

$$Q = Z^\top \bar{\Omega}^{-1} Z = (Y - \xi)^\top \Omega^{-1}(Y - \xi) \sim d \times F(d, v), \qquad (6.30)$$

which extends (4.13) to the d-dimensional case.

6.2.3 Applications of canonical form to the ST distribution

In the ST case, the terms ψ_m required by (6.20) and (6.21) are given by (4.14) on p. 103. Therefore, Mardia's coefficients take the form

$$\gamma_{1,d}^M = (\gamma_1^*)^2 + 3(d-1)\frac{\mu_*^2}{(v-3)\sigma_*^2}, \qquad \text{if } v > 3, \qquad (6.31)$$

$$\gamma_{2,d}^M = \beta_2^* + (d^2-1)\frac{(v-2)}{(v-4)} + \frac{2(d-1)}{\sigma_*^2}\left[\frac{v}{v-4} - \frac{(v-1)\mu_*^2}{v-3}\right] - d(d+2),$$
$$\text{if } v > 4, \qquad (6.32)$$

where $\mu_* = b_v\delta_*$, $\sigma_*^2 = (v-2)^{-1}v - \mu_*^2$, δ_* is given in (5.38), and b_v is given in (4.15). The explicit expressions of γ_1^* and $\beta_2^* = \gamma_2^* + 3$ are obtained by evaluating (4.18) and (4.19) at $\delta = \delta^*$.

The next statement is the analogue for the ST distribution of Proposition 5.14 for the SN. Similarly to the earlier result, one implication is the alignment of the location parameter, the mode and the mean, when this exists.

Proposition 6.2 *The unique mode of the* $ST_d(\xi, \Omega, \alpha, v)$ *distribution is*

$$M_0 = \xi + \frac{m_0^*}{\alpha_*}\omega\bar{\Omega}\alpha = \xi + \frac{m_0^*}{\delta_*}\omega\delta,$$

where δ is as in (5.11), α_* as in (5.37) and $m_0^* \in \mathbb{R}$ is the unique solution of the equation

$$x(v+d)^{1/2}T(w(x); v+d) - v\alpha_*(v+x^2)^{-1/2}t(w(x); v+d) = 0$$

with $w(x) = \alpha_*x(v+d)^{1/2}(v+x^2)^{-1/2}$.

6.2.4 Bibliographic notes

As indicated in the bibliographic notes at the end of § 4.3.1, the multivariate ST distribution examined here was obtained by Branco and Dey (2001) as a special case of the skew-elliptical distributions. Their expression of the ST density was, however, stated in the form (6.13). Expression (6.23) was obtained by Azzalini and Capitanio (2003) and Gupta (2003), independently from each other, starting from representation (6.22). These papers contain also additional properties, such as moments and distribution of quadratic forms. The proof of Proposition 6.2 is given by Capitanio (2012).

Additional results on the ST distribution theory are given by Kim and Mallick (2003), specifically on the moments of the distribution and those of its quadratic forms. Notice that their parameterization uses the same symbols employed here but is slightly different. One of their results is an expression for the Mardia coefficient of multivariate kurtosis; its amended version is equivalent to (6.32), although of quite different appearance.

Soriani (2007) adapts the procedure of Complement 5.2 to obtain regions of given probability for the bivariate ST distribution.

6.2.5 Statistical aspects

The statistical aspects of the multivariate ST distribution are qualitatively much the same as in the univariate context, as discussed in § 4.3.2 and § 4.3.3, but the technical side becomes substantially more intricate, leading to complex expressions which would take much space to replicate here. Therefore we only provide a general discussion, referring to specialized publications for the missing expressions.

The direct parameter set for the simple sample case is represented by $\theta^{\mathrm{DP}} = (\xi^{\top}, \mathrm{vech}(\Omega)^{\top}, \alpha^{\top}, \nu)^{\top}$. In the multivariate linear regression setting of (5.51), the location parameter ξ is replaced by $\mathrm{vec}(\beta)$. In some cases, we may want to regard ν as known, hence reducing the size of θ^{DP} by one. An instance of this type is represented by the skew-Cauchy distribution, where we set $\nu = 1$.

Under independence of a set of observations $y_1, \ldots, y_n$, the log-likelihood function of θ^{DP} is

$$\log L = \mathrm{constant}$$
$$+ n \left[\log \Gamma((\nu + d)/2) - (d/2) \log \nu - \log \Gamma(\nu/2) - \tfrac{1}{2} \log \det(\Omega) \right]$$
$$+ \sum_{i=1}^{n} \left[-\frac{\nu + d}{2} \left(1 + \frac{z_i^{\top} \bar{\Omega}^{-1} z_i}{\nu} \right) + \log T \left(\alpha^{\top} z_i \sqrt{\frac{\nu + d}{\nu + Q(z_i)}}; \nu + d \right) \right],$$
$$(6.33)$$

where $z_i = \omega^{-1}(y_i - \xi_i)$, $Q(z_i) = z_i^{\top} \bar{\Omega}^{-1} z_i$ and ξ_i is either a constant ξ or of the form (5.51) in the linear regression case.

Maximization of $\log L$ can only be tackled in a numerical form, either via direct optimization of (6.33) or via some algorithm of the EM family. For the first approach, numerical search can be speeded up considerably by providing the optimization algorithm with the partial derivatives of $\log L$ given by Azzalini and Capitanio (2003) in an appendix of the full version of the paper. Numerical differentiation of this gradient, evaluated at the MLE $\hat{\theta}^{\mathrm{DP}}$, provides the DP observed information matrix. This is the route followed for the numerical work of the next section. An alternative direction, via some EM-type algorithm, has been developed by Lachos *et al.*

(2010b) as well as by various other authors, usually in some more general framework of the type summarized in § 8.2.1; so additional references are provided there.

In a technically impressive paper, Arellano-Valle (2010) computes the expected Fisher information of θ^{DP}, and proves that this matrix is non-singular at $\alpha = 0$ for all $\nu > 0$. Ley and Paindaveine (2010a) prove non-singularity via a different route, under the assumption $\nu > 2$.

For the reasons already discussed in the univariate case, we also consider a CP summary of the distribution with the following components: μ and Σ as given by (6.25) and (6.26), respectively, the d-vector γ_1 of measures of skewness computed component-wise from (4.18), and the Mardia coefficient of multivariate excess kurtosis $\gamma_{2,d}^M$, whose expression for the ST distribution is given in § 6.2.3. The dimensionality of the components of $(\mu, \Sigma, \gamma_1, \gamma_{2,d}^M)$ matches those of $(\xi, \Omega, \alpha, \nu)$.

Again, the set of CP quantities requires that $\nu > 4$; when this condition is violated, alternatives must be introduced. One option is to consider the multivariate version of the 'pseudo-CP' introduced in § 4.3.4; see Arellano-Valle and Azzalini (2013). An alternative route to circumvent the possible non-existence of moments, already mentioned in § 4.3.4 for the univariate case, is to work with quantile-based measures and their analogues in the multivariate context, derived from the idea of depth function. A formulation of this type has been put forward by Giorgi (2012).

6.2.6 A numerical illustration (continued)

Consider the wine data of the Grignolino cultivar, some of which have been used in § 4.3.4. Here we examine the variables chloride, glycerol and magnesium; hence $d = 3$ and $n = 71$. Figure 6.1 displays the scatter plot matrix for all pairs of variables, with the contour lines of the fitted bivariate distributions superimposed, computed using (6.29). The levels of the curves are chosen appropriately so that each curve surrounds a region with approximate probability at a given level, which is denoted by its label; the probability level of the outermost regions is 0.95.

In a regression setting where each observation has a different location parameter, a graphical representation like in Figure 6.1 would be inappropriate. However, a simple suitable modification is to plot instead the residuals of the regression model, superimposing the marginal distributions with the location parameter set to $\xi = 0$.

To assess the quality of the fitted distribution in Figure 6.1, we make use of Healy type graphical diagnostics similar to those described in Chapter 5,

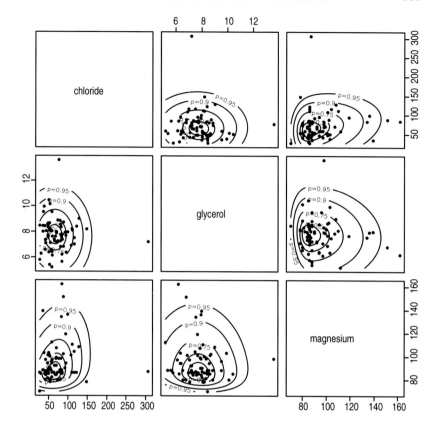

Figure 6.1 Wines data, three variables of Grignolino cultivar: scatter plot matrix of the observations with superimposed contour lines plots of the ST distribution fitted by maximum likelihood. The contour curves enclose regions of approximate probability indicated by their labels.

except that now the reference distribution of the sample quantities (5.55) is a scaled Snedecor's $F(d, v)$, from (6.30). In the present case, $d = 3$ and v is approximated by its MLE $\hat{v} = 3.4$.

The resulting graphical outcome is displayed in Figure 6.2 in the form of a QQ-plot and a PP-plot, in the left and right panels, respectively. These plots confirm the visual impression given by Figure 6.1 showing that, in most aspects, the contour lines accommodate the data scatter satisfactorily. There is one extremal point in the QQ-plot which deviates markedly from the ideal alignment line. Data inspection indicates that this point is the one which also appears isolated from the others in Figure 6.1, that is the one with highest value of chloride. This point is so far out from the others that it must be regarded as an outlier even for a long-tailed ST distribution with $\hat{\nu} = 3.4$. Notice, however, that the bivariate distributions are not shifted in its direction, but are placed around the main body of the data points.

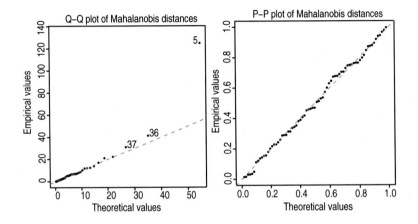

Figure 6.2 Wines data, three variables of Grignolino cultivar: QQ-plot (left panel) and PP-plot diagnostics (right-panel) of the fitted multivariate STdistribution.

To illustrate graphically the behaviour of the DP log-likelihood function, Figure 6.3 displays the profile deviance for some parameters of the fitted distribution in Figure 6.1. The left panel refers to (α_1, α_3), showing regular convex regions, without kinks at crossing $\alpha_1 = 0$, as happens in Figure 3.3(b). Similarly to Figure 4.9(a), the levels of these curves are chosen equal to appropriate percentage points of the χ_2^2 distribution, so that the enclosed regions represent confidence regions at the confidence levels indicated, up to an approximation. The right panel of Figure 6.3 displays the deviance function of $(\alpha_3, \log \nu)$, which again is largely regular. Transforming ν on the log-scale produces a more symmetric behaviour than on the original scale.

6.2.7 The multivariate extended skew-t family

We introduce an extension of the multivariate ST distribution via a similar construction of the extended multivariate SN. Recall from § 6.2.1 that the ST distribution can be generated by consideration of $(X_0|X_1 > 0)$ when (X_0, X_1) is a Student's t random variable with density $t_{d+1}(x; \Omega^*, \nu)$, where

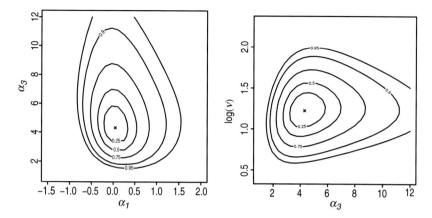

Figure 6.3 Wines data, three variables of Grignolino cultivar: profile deviance of (α_1, α_3) in the left panel and of $(\alpha_3, \log v)$ in the right panel. The crosses indicate the MLE point.

Ω^* is as in (6.12). For a given $\tau \in \mathbb{R}$, consider $Z \stackrel{d}{=} (X_0 | X_1 + \tau > 0)$, which we say to have a multivariate *extended skew-t distribution* (EST). Proceeding similarly to computation of (6.23), the density function of Z at $z \in \mathbb{R}^d$ turns out to be

$$f_Z(z) = \frac{1}{T(\tau; v)}\, t_d(z; \bar{\Omega}, v)\, T\left((\alpha_0 + \alpha^\top z)\sqrt{\frac{v + d}{v + Q(z)}}; v + d\right), \qquad (6.34)$$

where $Q(z) = z^\top \bar{\Omega}^{-1} z$, similarly to (6.23), and α_0 is as in (5.60). The parameter set is the same as (6.23) plus τ.

As usual, location and scale parameters must be introduced in practical work, in the form $Y = \xi + \omega Z$, as in (5.2). In this case, we use the notation $Y \sim \mathrm{ST}_d(\xi, \Omega, \alpha, v, \tau)$ where the presence of the final term indicates that we are dealing with an EST variable. The density of Y is computed as in (6.24).

Predictably, (6.34) combines aspects of the multivariate ST and of the multivariate ESN distribution, to which it reduces when $\tau = 0$ and when $v \to \infty$, respectively. We do not enter a detailed exploration of this distribution and only summarize the main findings of Adcock (2010) and Arellano-Valle and Genton (2010b) who, independently from each other, have put forward the EST distribution. These papers adopt different parameterizations; the one of Arellano-Valle and Genton (2010b) is, however, nearly the same as used here. The univariate EST has appeared in Jamalizadeh *et al.* (2009b).

Properties of closure under marginalization, conditioning and affine transformations hold for family (6.34), specifically as follows. If Y is partitioned as $Y^\top = (Y_1^\top, Y_2^\top)$ where the two components have dimension h and $d - h$, respectively, and correspondingly the parameters are partitioned as in (5.26) on p. 130, then marginally

$$Y_1 \sim ST_h(\xi_1, \Omega_{11}, \alpha_{1(2)}, \nu, \tau), \qquad (6.35)$$

where $\alpha_{1(2)}$ is equal to the ESN case, as given by (5.27), and conditionally

$$(Y_2|Y_1 = y_1) \sim ST_{d-h}\left(\xi_{2\cdot1}, q^2\, \Omega_{22\cdot1}, \alpha_{2\cdot1}, \nu + h, q^{-1}\tau_{2\cdot1}\right), \qquad (6.36)$$

where we have used quantities defined in the ESN case, by (5.64) and (5.66), and

$$q^2 = \frac{\nu + (y_1 - \xi_1)^\top \Omega_{11}^{-1}(y_1 - \xi_1)}{\nu + h}.$$

When $\alpha = 0$ in (6.34), we do not recover the classical Student's t distribution, but another elliptical distribution instead, while $\alpha = 0$ in (5.59) produces the Gaussian distribution. If both $\alpha = 0$ and $\tau = 0$, then we obtain the multivariate Student's t.

Besides the stochastic representation via conditioning, the multivariate EST distribution allows the following additive representation. If T_c is a univariate Student's t variate on ν d.f. truncated below $-\tau$ and T_0 is an independent d-dimensional Student's t with $\nu + 1$ degrees of freedom and scale matrix $\bar{\Omega} - \delta\delta^T$, where $\bar{\Omega}$ and δ are components of Ω^* as in (5.14), then

$$Z \stackrel{d}{=} \sqrt{\frac{\nu + T_c^2}{\nu + 1}}\, T_0 + \delta\, T_c$$

(Arellano-Valle and Genton, 2010b, Proposition 2). This representation is progressively more convenient for random number generation than that by conditioning as τ decreases to $-\infty$. Using this representation, Arellano-Valle and Genton have obtained expressions for univariate moments up to the fourth order and Mardia's coefficients of skewness and kurtosis.

Exploration of the formal properties of the likelihood function is technically complex. For the case $d = 1$, it can be shown that the information matrix is non-singular at $\alpha = 0 = \tau$ when ν is finite. Singularity occurs when $\nu \to \infty$ in addition to the above conditions; in this case, we return to the univariate ESN distribution.

Caution must be exercised if the EST distribution is fitted to data, for reasons similar to those discussed in § 3.3.2 for the univariate ESN distribution, which apply *a fortiori* in this more complex setting. In the EST case, an analogous formal analysis of the Fisher information is more difficult to pursue, especially in the multivariate case, but there is numerical evidence that also for the EST the profile log-likelihood function of τ is often nearly flat over a semi-infinite interval. Recall from § 3.3.2 that the instability in estimation of τ propagates its effect also on other parameters.

These problematic aspects in data fitting must not prevent the use of this distribution, for instance when a sufficiently high number of observations is available and also for other purposes. A very appropriate illustration of the latter type of use is represented by the method put forward by Lee *et al.* (2010) for data perturbation in connection with security problems of numerical databases; in this context, a standard problem is to avoid exact disclosure of confidential data in response to some query submitted to the database. One of the methods in use is to fit a multivariate distribution to the entire set of variables, comprising confidential and non-confidential variables, followed by computation of the conditional distribution of the confidential variables given the values taken by the others. From this conditional distribution random values are sampled to produce a perturbed version of the confidential variables. In this construction, the classical distribution in use for data fitting is the multivariate normal, but the ST can produce a better fit in this process. The subsequent conditioning step then leads to an EST distribution.

Another, and in a sense more substantial, application of the EST distribution has been presented by Marchenko and Genton (2012) who have considered an extension of the Heckman's formulation recalled in § 3.4.1. The original formulation is often criticized as being strongly dependent on the assumption of joint normality of the error terms $(\varepsilon_1, \varepsilon_2)$ appearing in (3.40). This is especially problematic considering that in the main application areas of this formulation, namely social statistics and economics studies, the distributions in play are often non-Gaussian, typically with longer tails. A sensible remedy is to relax the distributional assumption on $(\varepsilon_1, \varepsilon_2)$ by introducing a parametric class which allows us to regulate the tail thickness, and the bivariate Student's t is a quite natural candidate for the purpose. This route leads directly to the univariate EST as the distribution of $(Y|W > 0)$ with $\tau = w^\top \gamma$ and the other parameters as for the ST case.

A fortunate aspect of this application of the EST distribution, compared with to its plain use for data fitting, is that here there is a second source of information provided by the indicator variable $W^* = I_{(0,\infty)}(W)$. This supplements the EST likelihood with that associated with n Bernoulli trials having probability of success $T(w_1^\top \gamma), \ldots, T(w_n^\top \gamma)$, respectively. The score function for γ is not 0 at $\alpha = 0$, different from the score for τ of the EST, and also elsewhere the problem of flatness of the log-likelihood is reduced (personal communication of one of the authors), leading to more reliable inferences. Marchenko and Genton (2012) carried out a detailed simulation study which provides evidence of a clear improvement over the traditional Heckman's formulation.

Jamalizadeh *et al.* (2009b) developed a recursive relationship for the univariate EST distribution function, starting from the same function with lower d.f.; they also provide expressions for the initial values of the recursion. These results represent the extension to general τ of the results summarized in Complement 4.3 for the ST distribution, that is, for $\tau = 0$.

6.2.8 Some applications

The Black–Litterman technique allows us to construct financial portfolios under assumption of joint multivariate Gaussian distribution for a set of quantitative indicators. Meucci (2006) extends this technique by replacing the Gaussian assumption with an ST distribution.

The motivation for the already-quoted work of Adcock (2010) on the multivariate EST distribution came from quantitative finance. This paper can be viewed as a development of Adcock and Shutes (1999), and the financial optimality problems examined earlier were reconsidered in 2010 under the more flexible context provided by EST.

Thompson and Shen (2004) have examined a time series of hourly sea levels recorded over 81 years with the aim of evaluating the risk of coastal flooding. After removing the tide effect and other data preprocessing, the sequence of adjacent pairs of residuals is modelled as a bivariate ST distribution with time-dependent scale parameter to reflect seasonal variation. This model, recombined with the tide effect, is then used for evaluating the risk of flooding.

Ghizzoni *et al.* (2010; 2012) have employed the multivariate ST as a model for hydrological processes. Specifically, they have considered the joint distribution of river flow recorded at various gauging stations belonging to the same river basin, to assess risk of flooding in a comprehensive formulation. Two river basins have been considered and a d-dimensional ST distribution has been fitted to the joint river-flow distribution, in one case at $d=3$ gauging stations, in the other case at $d=18$ stations. As for the ability to fit the observed data, the ST outcome turned out to be comparable with that obtained with the more commonly used approach based on copulae, with some advantage in simplicity.

Many other applications have been considered, in a range of different areas. Quite a few of them are collected in the book edited by Genton

(2004). Since many of these developments intersect with advances in statistical methodology as well as use of probability distributions presented in Chapter 7, we defer their illustration to §8.2.

6.3 Complements

Complement 6.1 (When is the information matrix singular?)   The discussion in § 4.3.3 has shown that for the ST distribution with fixed d.f. the expected Fisher information for $\theta^{\mathrm{DP}} = (\xi, \omega, \alpha)$ is non-singular at $\alpha = 0$, while we had seen in Chapter 3 that a similar setting for the SN distribution leads to a singular information matrix, and the same happens in the more general setting of Problem 3.2. The question is then: in which distributions does this singularity arise?

For simplicity, we discuss the problem in the univariate case. Consider the family of one-dimensional density functions

$$f(x) = 2\omega^{-1} f_0(z) G(z; \alpha), \qquad z = \omega^{-1}(x - \xi), \qquad (6.37)$$

regulated by the parameters (ξ, ω, α), where f_0 is a density symmetric about 0 and $G(z; \alpha)$ satisfies (1.4) for any fixed values of α. We choose that the value $\alpha = 0$ corresponds to symmetry, hence $G(z; 0) = \frac{1}{2}$ for all $z \in \mathbb{R}$. Moreover, we make the mild assumption that $G'(z; 0) = 0$, where G' is the partial derivative of G with respect to z; we are then requiring that differentiation with respect to z and evaluation at $\alpha = 0$ are interchangeable. Families like the three-parameter SN distribution (2.3), distribution (2.53), when this is complemented with location and scale, and many other distributions satisfy this requirement. The same holds also for the asymmetric Subbotin distribution (4.5) and for the ST family when the tail-weight parameter ν is fixed. In addition, we make the regularity assumptions required to ensure that all derivatives and expectations involved in the following steps exist.

The score function for a single observation y evaluated at $\theta_0 = (\xi, \omega, 0)$ is

$$S(\theta_0) = \begin{pmatrix} \omega^{-1} h(z) \\ \omega^{-1} [z\,h(z) - 1] \\ 2\,\dot{G}(z) \end{pmatrix},$$

where $z = \omega^{-1}(y - \xi)$, $h(z) = -f_0'(z)/f_0(z)$ and $\dot{G}$ denotes the partial derivative of G with respect to α, evaluated at $\alpha = 0$. The expected Fisher information can be computed as the variance matrix of $S(\theta_0)$, leading to

$$\mathcal{I}(\theta_0) = \begin{pmatrix} \mathcal{I}(\theta_0)_{11} & 0 & \mathcal{I}(\theta_0)_{13} \\ 0 & \mathcal{I}(\theta_0)_{22} & 0 \\ \mathcal{I}(\theta_0)_{13} & 0 & \mathcal{I}(\theta_0)_{33} \end{pmatrix}$$

where

$$\mathcal{I}(\theta_0)_{11} = \omega^{-2}\,\mathbb{E}\{h(Z)^2\}, \qquad \mathcal{I}(\theta_0)_{13} = 2\omega^{-1}\,\mathbb{E}\{\dot{G}(Z)\,h(Z)\},$$
$$\mathcal{I}(\theta_0)_{22} = \omega^{-2}\,\mathbb{E}\{[Z\,h(Z) - 1]^2\}, \quad \mathcal{I}(\theta_0)_{33} = 4\,\mathbb{E}\{\dot{G}(Z)^2\}$$

while the null terms in $\mathcal{I}(\theta_0)$ occur because the first and last components of $S(\theta_0)$ are odd functions of z, the second component is an even function, and we integrate their products with respect to an even density.

Singularity of $\mathcal{I}(\theta_0)$ can only occur when the submatrix obtained by eliminating the second row and column has null determinant, that is when

$$\mathbb{E}\{h(Z)^2\}\,\mathbb{E}\{\dot{G}(Z)^2\} = \mathbb{E}\{\dot{G}(Z)\,h(Z)\}^2 .$$

The Cauchy–Schwarz inequality ensures that this equality holds if and only if $h(z) = a\,\dot{G}(z)$ for some constant a, with probability 1. On replacing $h(z)$ by its definition and solving the ensuing differential equation, we conclude that f_0 in (6.37) must be of the exponential family type

$$f_0(z) = c\,\exp\{-a\,G^*(z)\}, \qquad z \in \mathbb{R}, \tag{6.38}$$

where G^* is a primitive of $\dot{G}$ and c is a normalizing constant. This f_0 is of exponential type with natural parameter $-a$ and sufficient statistics $G^*(z)$.

For instance, the popular case with $G(z; \alpha) = G_0(\alpha z)$, where G_0 satisfies (1.1), lends $\dot{G}(z) = G_0'(0)\,z$. Unless $G_0'(0) = 0$, singularity occurs when $f_0(z)$ is the $N(0, \sigma^2)$ density for some σ^2.

Bibliographic notes

The above discussion is based on work of Hallin and Ley (2012), where the argument is extended further to the multivariate case. The earlier paper by Ley and Paindaveine (2010b), which deals with some related problems in a multivariate setting, includes among others the result that stationarity of the profile log-likelihood at $\alpha = 0$ occurs only with a normal base density f_0 when the perturbation factor is of type $G_0\{\alpha^\top(x - \xi)\}$. Singularity of the information matrix for a class of univariate distributions with a normal base density, mentioned in connection with Problem 3.2, has been shown by Pewsey (2006b).

Complement 6.2 (Quasi-concavity)   For a density $f(x)$ on $\mathbb{R}^d$, consider the regions enclosed by the contour lines, that is the sets of type

$$R(c) = \{x : f(x) \geq c, x \in \mathbb{R}^d\}$$

for any $c > 0$. If $R(c)$ is a convex set for all $c > 0$, density f is said to be quasi-concave. For $d = 1$ the condition of quasi-concavity coincides with that of unimodality, but for $d > 1$ the two concepts separate, quasi-concavity being a stronger condition than unimodality. A graphical illustration of bivariate unimodal densities which are not quasi-concave is provided by the top plots of Figure 1.1 on p. 5.

The $R(c)$ sets of bivariate ST densities in Figure 6.1 appear instead to be all convex. Can we confirm formally this graphical appearance, possibly for all ST densities? More generally, can we say when a skew-elliptical distribution is quasi-concave?

In Chapter 5 we have seen that the SN density is log-concave, and this implies quasi-concavity. For skew-elliptical distributions, log-concavity cannot hold in general, since it does not hold for the subset of symmetric densities. For instance, the symmetric t density is not log-concave, for any choice of d. Nevertheless, quasi-concavity of the t density holds, as for all elliptical densities, by their very definition.

Therefore a specific treatment is required. This has been tackled by Azzalini and Regoli (2012a) by making use of the notion of s-concavity and other facts presented by Dharmadhikari and Joag-dev (1988). Since the development is fairly technical, we only summarize the main facts.

If $s < 0$, we say that f is s-concave if f^s is convex; if $s > 0$, the requirement is that f^s is concave; limiting cases are handled by continuity. If $s = 1$, then s-concavity represents ordinary concavity of f; the case $s = 0$ corresponds to log-concavity; $s = -\infty$ corresponds to quasi-concavity. The notion of s-concavity constitutes a graded form of concavity, increasingly stringent as s increases along the real line, that is, if f is s-concave, it is r-concave for all $r < s$.

The essence of the answer to the above-raised question is then as follows. If the density generator $\tilde{p}$ of X in (6.12) is decreasing and s-concave with $s \geq -1$, the density (6.13) of Z is s_1-concave with $s_1 = s/(1+s)$, hence also quasi-concave. In essence, quasi-concavity holds for a skew-elliptical density if the parent $(d+1)$-dimensional density of X satisfies this slightly stronger form of s-concavity. Some condition of this sort cannot be avoided completely, since one can construct examples of skew-elliptical densities which are not quasi-concave even starting from a quasi-concave elliptical density for X.

Using the above general result, we can examine various special families of the class (6.13). An important case is represented by the Pearson type VII family with density generator $\tilde{p}$ given by (6.8). This generator is decreasing and s-concave with $s = -1/M$. Since $s \geq -1$, the above general result ensures s_1-concavity with $s_1 = -1/(M-1)$. In particular, for the ST distribution on ν d.f., which corresponds to $M = (d+\nu+1)/2$, we conclude that it is s_1-concave with $s_1 = -2/(d+\nu-1)$. Hence the multivariate ST density is quasi-concave for all d and all ν. This implies that the regions delimited by contour lines are convex and consequently that the density is unimodal.

Complement 6.3 (Skew-Cauchy distributions) There are various possible formulations for a multivariate skew-Cauchy distribution.

⋄ Type I. On setting $\nu = 1$ in (6.23), we obtain a form of multivariate skew-Cauchy distribution, which is the direct extension of the univariate distribution of Complement 4.2. In a similar logic, we can set $\nu = 1$ in (6.34) to produce an extended version of this distribution.

⋄ Type II. An alternative type of skew-Cauchy distribution has been examined by Arnold and Beaver (2000b). Given independent standard Cauchy variables $Z_1, \ldots, Z_d, U$, consider the distribution of $Z = (Z_1, \ldots, Z_d)^\top$ conditionally on $\alpha_0 + \alpha^\top Z > U$. The key difference from a type I construction is that here the conditioning mechanism operates on a set of $d+1$ independent Cauchy variables, which jointly do not have an elliptical distribution, by Proposition 6.1(e). In this case the modulating factor is in the 'extended form' (1.26), and the normalizing constant must be computed afresh. The density at $x = (x_1, \ldots, x_d)^\top \in \mathbb{R}^d$ turns out to be

$$\prod_{j=1}^{d} h(x_j) \frac{H(\alpha_0 + \alpha^\top x)}{H[\alpha_0/(1 + \sum_j |\alpha_j|)]}, \tag{6.39}$$

where

$$h(z) = \frac{1}{\pi(1+z^2)}, \qquad H(z) = \frac{1}{2} + \frac{\arctan z}{\pi} \tag{6.40}$$

denote the standard Cauchy density and distribution function at $z \in \mathbb{R}$.

For applied work Arnold and Beaver introduce location and scale factors via the transformation $Y = \xi + \Omega^{1/2}Z$, where Z denotes a variable with density (6.39), $\xi \in \mathbb{R}^d$ and $\Omega^{1/2}$ is a $d \times d$ matrix.

⋄ Type III: see Problem 6.6.

Complement 6.4 (Another skew-elliptical family) Sahu *et al.* (2003)

have proposed an alternative form of skew-elliptical family, whose key difference from distributions like (6.11) and (6.12) is that the formulation involves as many latent variables as those observed. To start with, consider the $2d$-dimensional variable

$$\begin{pmatrix} \varepsilon \\ Z_1 \end{pmatrix} \sim \mathrm{EC}_{2d}\left(\begin{pmatrix} \xi \\ 0 \end{pmatrix}, \begin{pmatrix} \Sigma & 0 \\ 0 & I_d \end{pmatrix}, \tilde{p}\right)$$

where Σ is a $d \times d$ symmetric positive-definite matrix and $\xi \in \mathbb{R}^d$. Next define $Z_0 = \varepsilon + \Lambda Z_1$ where $\Lambda = \mathrm{diag}(\lambda_1, \ldots, \lambda_d)$ has non-zero diagonal elements, and finally consider the distribution of $Z \overset{\mathrm{d}}{=} (Z_0|Z_1 > 0)$ where $Z_1 > 0$ means that all components of Z_1 are positive. To find the distribution Z, consider the joint distribution

$$\begin{pmatrix} Z_0 \\ Z_1 \end{pmatrix} \sim \mathrm{EC}_{2d}\left(\begin{pmatrix} \xi \\ 0 \end{pmatrix}, \begin{pmatrix} \Sigma + \Lambda^2 & \Lambda \\ \Lambda & I_d \end{pmatrix}, \tilde{p}\right)$$

using (6.3). We then apply the same argument leading to (1.28), where in the present case we set $m = d$ and C is represented by the orthant of $\mathbb{R}^d$ with positive coordinates. Since $Z_1 \sim \mathrm{EC}_d(0, I_d, \tilde{p}_d)$, then $\mathbb{P}\{C\} = \mathbb{P}\{Z_1 > 0\} = 2^{-d}$. The other ingredient of (1.28) is the distribution of $(Z_1|Z_0 = z)$, which can be computed from (6.7), leading to

$$(Z_1|Z_0 = z) \sim \mathrm{EC}_d\left(\Lambda(\Sigma + \Lambda^2)^{-1}z_0, I_d - \Lambda(\Sigma + \Lambda^2)^{-1}\Lambda, \tilde{p}_{Q(z_0)}\right), \quad (6.41)$$

where $z_0 = z - \xi$, $Q(z_0) = z_0^\top(\Sigma + \Lambda^2)^{-1}z_0$ and $\tilde{p}_{Q(z_0)}$ is the density generator of the conditional distribution. Combining these terms, we write the density of Z at $z \in \mathbb{R}^d$ as

$$f_z(z) = 2^d\, p_d(z; \xi, \Sigma + \Lambda^2)\, \mathbb{P}\{Z_1 > 0|Z_0 = z\} \quad (6.42)$$

where p_d is of type (6.2). In general, the final factor of (6.42) must be evaluated by numerical integration of (6.41).

When $d = 1$ this formulation coincides with that of (6.12), up to a change of parameterization; differences appear for $d > 1$. For instance, when the parent elliptical variable is normal, the final factor in (6.42) becomes $\mathbb{P}\{U > 0\}$ where $U \sim N_d(\Lambda(\Sigma + \Lambda^2)^{-1}(z - \xi); I_d - \Lambda(\Sigma + \Lambda^2)^{-1}\Lambda)$, leading to the density function

$$f_z(z) = 2^d\, \varphi_d(z - \xi; \Sigma + \Lambda^2)$$
$$\times \Phi_d\left(\Lambda(\Sigma + \Lambda^2)^{-1}(z - \xi); I_d - \Lambda(\Sigma + \Lambda^2)^{-1}\Lambda)\right). \quad (6.43)$$

This density represents a form of skew-normal distribution alternative to (5.1), if $d > 1$. They share the same base density but differ in the modulation factor. A visible difference is that, if Σ is diagonal, (6.43) factorizes as the product of d univariate skew-normal densities, hence including d factors of type $\Phi(\cdot)$, while (5.1) has one such factor only. Families (6.43) and (5.1) share the same number of parameter components, $2d+d(d+1)/2$, and none of them is a superset of the other family.

In a similar fashion, an alternative form of multivariate ST distribution can be considered; see Sahu *et al.* (2003, Section 4). Similarly to the SEC family discussed earlier, the SN and ST forms are two instances of this formulation which have received attention from the subsequent literature recalled below.

As mentioned above, the final factor of (6.42) is usually hard to compute. Since this term is not required for practical data fitting via the MCMC technique in a Bayesian context, it is natural that this formulation has received much interest in the Bayesian literature. Besides the treatment of regression models with these error distributions by Sahu *et al.* (2003, Sections 5–6), other instances are the work by Sahu and Dey (2004) on multivariate frailty models, by Tchumtchoua and Dey (2007) on stochastic frontier models, and by De la Cruz (2008) on non-linear regression for longitudinal data. However, these distributions can also be employed in the classical framework, as demonstrated by Lin (2010) and Lin and Lin (2011), who have presented EM-type algorithms for MLE computation where the E-step is pursued via simulation methods. Lee and McLachlan (2012) have developed classical and faster EM algorithms for this setting, as well as for some distributions discussed in earlier sections.

Complement 6.5 (Extreme values and tail dependence)   In the theory of extreme value distributions, an important concept is that of tail dependence for a bivariate variable $X = (X_1, X_2)$. If F_1 and F_2 denote the distribution functions of X_1 and X_2, the commonly employed coefficients to express dependence between extreme values of the components are

$$\lambda_U = \lim_{u \to 1} \mathbb{P}\left\{X_1 > F_1^{-1}(u) | X_2 > F_2^{-1}(u)\right\}, \tag{6.44}$$

$$\lambda_L = \lim_{u \to 0} \mathbb{P}\left\{X_1 \le F_1^{-1}(u) | X_2 \le F_2^{-1}(u)\right\}, \tag{6.45}$$

which refer to the upper and the lower tail, respectively. If λ_U exists and is positive, X is said to have positive upper tail dependence; if $\lambda_U = 0$, we say that X_1 and X_2 are upper tail independent. The meaning of λ_L is the same with respect to the lower tail. It is immediate to check that for a centrally symmetric distribution the two measures coincide. An example of distributions with positive tail dependence is the Student's t, while the bivariate normal has independent tails.

An interesting feature of the bivariate ST distribution is to allow for different grades of dependence in the lower and upper tail, as shown by Fung and Seneta (2010) and Padoan (2011) in independent work. Specifically, if $X \sim ST_2(0, \bar{\Omega}, \alpha, v)$ with the off-diagonal element of $\bar{\Omega}$ equal to ρ, they obtain that

$$\lambda_U = \mathbb{P}\left\{Y_1 > \frac{(a_1^{1/v} - \rho)\sqrt{v+1}}{\sqrt{1-\rho^2}}\right\} + \mathbb{P}\left\{Y_2 > \frac{(a_2^{1/v} - \rho)\sqrt{v+1}}{\sqrt{1-\rho^2}}\right\} \qquad (6.46)$$

where the marginal distributions of

$$Y_1 \sim ST\left(0, 1, \alpha_1 \sqrt{1-\rho^2}, v+1, \alpha_{2(1)}\sqrt{v+1}\right),$$
$$Y_2 \sim ST\left(0, 1, \alpha_2 \sqrt{1-\rho^2}, v+1, \alpha_{1(2)}\sqrt{v+1}\right)$$

have been computed from the expression (6.36) for conditional distributions, $\alpha_{1(2)}$ and $\alpha_{2(1)}$ are given by (5.27) and finally

$$a_1 = \frac{T(\alpha_{2(1)}\sqrt{v+1}; v+1)}{T(\alpha_{1(2)}\sqrt{v+1}; v+1)}, \qquad a_2 = \frac{T(\alpha_{1(2)}\sqrt{v+1}; v+1)}{T(\alpha_{2(1)}\sqrt{v+1}; v+1)}.$$

These expressions of a_1 and b_2 are slightly different from those of Padoan (2011) because of a typographical error in the quoted degrees of freedom of the Student's distribution function.

Since $-X \sim ST_2(0, \bar{\Omega}, -\alpha, v)$, the value λ_L for X coincides with that of λ_U for $-X$. In practice, this amounts to reversing the signs in the above expressions of $\alpha_{1(2)}, \alpha_{2(1)}, \alpha_1$ and α_2.

When $\alpha = 0$, the measures of tail dependence reduce to those of the regular Student's t. When $v \to \infty$, both measures converge to 0, implying that for the SN distribution there is tail independence, like for the normal.

Lysenko *et al.* (2009) obtain sufficient conditions for tail independence for distributions of type (1.3) when the base density is multivariate normal. In the SN case, these conditions allow us to prove tail independence within a certain subset of the parameter space.

Padoan (2011) provides additional results for extreme value theory, specifically the limit distribution of component-wise maxima of a sequence of independent random vectors with a common multivariate ST distribution.

Problems

6.1 Prove the statement of § 6.1.3 that an additive representation of type (5.19) holds also for skew-elliptical distributions (Azzalini and Capitanio, 2003; Fang, 2003).

6.2 Prove the statement of § 6.1.3 that a representation through minima and maxima similar to (5.25) holds also for skew-elliptical distributions.

6.3 If X in (6.12) has a Student's t density $t_{d+1}(x; \Omega^*, \nu)$ prove that the conditional density of X_0 given that $X_1 > 0$ is (6.23) (Azzalini and Capitanio, 2003, Proposition 4).

6.4 Check that the density of the variable Z defined by (6.22) is (6.23).

6.5 Confirm that (6.39) is a proper density function on $\mathbb{R}^d$, that is, show that $H[\alpha_0/(1 + \sum_j |\alpha_j|)]$ represents the appropriate normalizing constant. Show also that the m-dimensional distribution obtained by considering m components ($m \le d$) of (6.39) is still of the same type (Arnold and Beaver, 2000b).

6.6 Start from the variables $Z_1, \ldots, Z_d, U$ introduced in Complement 6.3 and define $W_j = Z_j/|U|$, for $j = 1, \ldots, d$. Show that $W = (W_1, \ldots, W_d)$ has density at $x \in \mathbb{R}^d$ equal to

$$\frac{\Gamma(\tfrac{1}{2}(d + 1))}{\pi^{(d+1)/2}} \frac{1}{(1 + x^\top x)^{(d+1)/2}} \frac{H(\alpha_0 + \alpha^\top x)}{H[\alpha_0/(1 + \sqrt{\alpha^\top \alpha})]},$$

where H is as in (6.40). This lends another form of multivariate skew-Cauchy distribution (Arnold and Beaver, 2000b).

6.7 Consider a skew-elliptical distribution of type (6.13) generated from (6.12) when the density generator is $\tilde{p} = (1 - x)^\nu$ for $x \in (0, 1)$, corresponding to a Pearson type II distribution. By using the results summarized in Complement 6.2, show that this skew-elliptical density is log-concave.

6.8 In the notation of § 6.2.7, confirm that the EST density (6.34) is the distribution of $Z \stackrel{d}{=} (X_0 | X_1 + \tau > 0)$.

6.9 Confirm the marginal and conditional EST distributions as given by (6.35) and (6.36) (Arellano-Valle and Genton, 2010b, up to a change in the parameterization).

6.10 Confirm that under normality (6.42) takes the form (6.43). Also, show that, if a variable Z has this distribution, then

$$\mathbb{E}\{Z\} = \xi + b\lambda, \qquad \text{var}\{Z\} = \Sigma + (1 - b^2)\Lambda^2$$

where $b = \sqrt{2/\pi}$ (Sahu et al., 2003). *Note:* While the mean is analogous to (5.31) on setting $\lambda = \omega\delta$, the variance is of a different form compared with (5.32).

6.11 As recalled in § 6.1.1, the multivariate slash distribution refers to a variable of type (6.10) when $V^{q/2} \sim U(0, 1)$ for some $q > 0$. Define the density of the skew-slash distribution similarly by assuming $W \sim$ $SN_d(0, \Omega, \alpha)$. Show that the same density is obtained starting from a $(d+1)$-dimensional slash variable X and applying the conditioning mechanism of (6.14). Compute the mean and variance matrix of the distribution (Wang and Genton, 2006).

6.12 Verify expressions (6.20) and (6.21) (Capitanio, 2012).

6.13 Prove Proposition 6.2 (Capitanio, 2012).

7

Further extensions and other directions

In the remaining two chapters of this book we consider some more specialized topics. The enormous number of directions which have been explored
prevent, however, any attempt at a detailed discussion within the targeted
area. Consequently, we adopt a quite different style of exposition compared
with previous chapters: from now on, we aim to present only the key concepts of the various formulations and their interconnections, referring more
extensively to the original sources in the literature for a detailed treatment.
Broadly speaking, this chapter focuses more on probabilistic aspects, the
next chapter on statistical and applied work.

7.1 Use of multiple latent variables

7.1.1 General remarks

In Chapters 2 to 6 we dealt almost exclusively with distributions of type
(1.2), or of its slight extension (1.26), closely associated with a selection
mechanism which involves one latent variable; see (1.8) and (1.11). For
the more important families of distributions, an additional type of genesis
exists, based on an additive form of representation, of type (5.19), which
again involves an auxiliary variable. Irrespective of the stochastic representation which one prefers to think of as the underlying mechanism, the
effect of this additional variable is to introduce a factor of type $G_0\{w(x)\}$ or
$G_0\{\alpha_0 + w(x)\}$ which modulates the base density, where G_0 is a univariate
distribution function.

The next stage of development is to consider a selection mechanism
which involves a number m, say, of latent variables, to reflect a more complex form of selection than just exceeding a certain threshold of one variable. A formulation of this type does not fall within the scheme (1.2) or
(1.26), but can be accommodated within (1.28). The operational difficulty

here is due to the computation of the two integrals appearing in the final factor of (1.28), especially so when the conditioning set C is awkward.

However, there exist some settings which involve tractable computations. In the reasonably simplified case when C can be expressed via a set of inequalities of type $c_k^\top X_1 > c_k'$ where c_k is an m-vector of constants and c_k' is a scalar ($k = 1, \ldots, m$), and closure under linear transformations holds for the family of distributions, the constraints can be expressed as $X_k' > c_k'$, where $X_k' = c_k^\top X_1$. In these circumstances, since we are transforming latent variables into other latent variables, C can be reduced to a rectangular region. This simplification does not entail a loss of generality, provided there is no qualitative structure of the latent variables, such as independence, that we want to enforce, since it would typically disappear after transformation.

A point of interest is whether equivalent stochastic representations exist for a distribution generated by the selection mechanism (1.28), notably a representation of additive type. However, in practice, the question cannot be examined without introducing some specification on the distribution of the elements, similarly to previous chapters.

7.1.2 Normal parent components

Extend (5.14) by assuming that the component X_1 is m-dimensional; more specifically, consider the random variable

$$X = \begin{pmatrix} X_0 \\ X_1 \end{pmatrix} \sim N_{d+m}(0, \Omega^*), \qquad \Omega^* = \begin{pmatrix} \bar{\Omega} & \Delta \\ \Delta^\top & \Gamma \end{pmatrix}, \qquad (7.1)$$

where Ω^* is still a full-rank correlation matrix. Define Z to be a random variable with distribution $(X_0 | X_1 + \tau > 0)$, where $\tau \in \mathbb{R}^m$ and the notation $X_1 + \tau > 0$ means that the inequality sign must hold for all m components. Under distribution (1.28), the computations involved are conceptually simple, since the conditional distribution of $(X_0 | X_1 = x_1)$ is still of multinormal type, with well-known expressions for the parameters. For the transformed variable $Y = \xi + \omega Z$, we obtain that the density is

$$f_Y(x) = \varphi_d(x - \xi; \Omega) \frac{\Phi_m\left\{\tau + \Delta^\top \bar{\Omega}^{-1} \omega^{-1}(x - \xi); \Gamma - \Delta^\top \bar{\Omega}^{-1}\Delta\right\}}{\Phi_m(\tau; \Gamma)} \qquad (7.2)$$

at $x \in \mathbb{R}^d$; here ξ and ω are as in (5.2) and $\Omega = \omega \bar{\Omega} \omega$. If $m = 1$, it is immediate that (7.2) reduces to the ESN density (5.61) and if, in addition, $\tau = 0$ we obtain the SN density (5.3). For reasons explained at the end of this section, we shall refer to (7.2) as a SUN distribution, and write

$$Y \sim SUN_{d,m}(\xi, \Omega, \Delta, \tau, \Gamma)$$

under the condition that Ω^* in (7.1) is non-singular. If $m = 1$, this corresponds to parameterizing a multivariate SN distribution via (Ω, δ), a choice which involves in fact a restriction on the components; see Problem 5.4.

A direct proof that (7.2) integrates to 1 can be obtained by the following simple extension of Lemma 5.2.

Lemma 7.1   *If $U \sim N_p(0, \Sigma)$ then*

$$\mathbb{E}\left\{\Phi_q(H^\top U + k; \Psi)\right\} = \Phi_q(k; \Psi + H^\top \Sigma H) \tag{7.3}$$

for any choice of the vector $k \in \mathbb{R}^q$, the $p \times q$ matrix H and the $q \times q$ symmetric positive-definite matrix Ψ.

Proof   If π denotes the left-hand term of (7.3) and $W \sim N_q(0, \Psi)$, independent of U, then

$$\pi = \mathbb{E}\{\mathbb{P}\{W \le H^\top U + k | U\}\} = \mathbb{P}\{W - H^\top U \le k\} = \Phi_q(k; \Psi + H^\top \Sigma H)$$

where the inequalities are intended to hold component-wise.   QED

Also in this case, an additive type of representation exists, based on independent variables $U_0 \sim N_d(0, \bar{\Psi}_\Delta)$ and $U_{1,-\tau}$, which is obtained by the component-wise truncation below $-\tau$ of a variate $U_1 \sim N_m(0, \Gamma)$. If $\bar{\Psi}_\Delta = \bar{\Omega} - \Delta\Gamma^{-1}\Delta^\top$, it can be shown that the distribution of

$$\xi + \omega \left(U_0 + \Delta\Gamma^{-1} U_{1,-\tau}\right) \tag{7.4}$$

is (7.2). The correspondence between the variables X_0, X_1 in (7.1) and the variables U_0, U_1 can be established by orthogonalization, like in (5.23).

The moment generating function of Y can be computed by direct evaluation of $\mathbb{E}\{\exp(t^\top Y)\}$, following the scheme of Lemma 5.3 and using (7.3). The result is

$$M(t) = \exp\left(\xi^\top t + \tfrac{1}{2}t^\top \Omega t\right) \frac{\Phi_m(\tau + \Delta^\top \omega t; \Gamma)}{\Phi_m(\tau; \Gamma)}, \qquad t \in \mathbb{R}^d. \tag{7.5}$$

From here it is immediate to obtain several properties of the family (7.2). One of these is closure under marginalization and affine transformations, a fact which could however be established by the genesis of the distribution. Specifically, if we partition $Y = (Y_1^\top, Y_2^\top)^\top$ where the components Y_1 and Y_2 have dimension d_1 and d_2, respectively, with $d_1 + d_2 = d$, then

$$Y_1 \sim SUN_{d_1, m}(\xi_1, \Omega_{11}, \Delta_1, \tau, \Gamma) \tag{7.6}$$

where ξ_1 and Ω_{11} are as in (5.26) and Δ_1 is formed by the first d_1 rows of Δ.

If a is a p-vector and A is a full-rank $d \times p$ matrix, then

$$a + A^\top Y \sim \mathrm{SUN}_{p,m}(a + A^\top \xi, A^\top \Omega A, \Delta_A, \tau, \Gamma),$$
$$\Delta_A = ((A^\top \Omega A) \odot I_p)^{-1/2} A^\top \omega \Delta.$$

From the ratio of the joint and the marginal distribution, one obtains after some algebra the conditional distribution

$$(Y_2 | Y_1 = y_1) \sim \mathrm{SUN}_{d_2,m}(\xi_{2\cdot1}, \Omega_{22\cdot1}, \Delta_{2\cdot1}, \tau_{2\cdot1}, \Gamma_{2\cdot1}) \qquad (7.7)$$

where $\xi_{2\cdot1}$ and $\Omega_{22\cdot1}$ are as in (5.64) and

$$
\begin{aligned}
\tau_{2\cdot1} &= \tau_2 + \Delta_1^\top \bar{\Omega}_{11}^{-1} \omega_1^{-1}(y_1 - \xi_1), \\
\Delta_{2\cdot1} &= \Delta_2 - \bar{\Omega}_{21} \bar{\Omega}_{11}^{-1} \Delta_1, \\
\Gamma_{2\cdot1} &= \Gamma - \Delta_1^\top \bar{\Omega}_{11}^{-1} \Delta_1,
\end{aligned}
$$

which are based on (5.28) and other quantities as in § 5.3.2.

A little computation using (7.5) lends closure under convolution: if Y_1 and Y_2 are independent variables of type (7.2) with dimensional indices (d, m_1) and (d, m_2), respectively, then from (7.5) we see that $Y = Y_1 + Y_2$ is of SUN type with indices $(d, m_1 + m_2)$ and parameters

$$
\xi = \xi_1 + \xi_2, \quad \Omega = \Omega_1 + \Omega_2, \quad \omega = (\omega_1^2 + \omega_2^2)^{1/2},
$$
$$
\Gamma = \begin{pmatrix} \Gamma_1 & 0 \\ 0 & \Gamma_2 \end{pmatrix}, \quad \tau = \begin{pmatrix} \tau_1 \\ \tau_2 \end{pmatrix}, \quad \Delta = (\omega^{-1}\omega_1 \Delta_1 \quad \omega^{-1}\omega_2 \Delta_2), \qquad (7.8)
$$

in an obvious notation.

The less appealing side of the SUN formulation is represented by the increased complexity, in various aspects. One is that the mere evaluation of the density (7.2) becomes cumbersome when m is not small, because of the Φ_m factor. In principle, computation of moments of any order is possible from (7.5) but intricate in practice beyond the first-order moment.

Also, from the statistical viewpoint, it is often the case that a construction involving a high number of latent variables may be troublesome to estimate reliably, unless a vast amount of data is available.

These problems can be reduced by introducing some restrictions on the parameters. A major simplification occurs if Γ is diagonal, leading to simplified expressions of the mean and variance. For instance, this is the case for the formulation summarized in Complement 6.4, where additionally it is required that $d = m$ and Δ is diagonal. However, if one wants to link this type of formulation to subject-matter motivations, the assumption of joint independence among the latent variables is clearly quite strong.

Bibliographic notes

The restricted formulation of Complement 6.4, just recalled, has been considered by Sahu *et al.* (2003). The unrestricted formulation has been studied in a series of papers by González-Farías *et al.* (2004a), Gupta *et al.* (2004) and González-Farías *et al.* (2004b), under the heading *closed skew-normal* distribution, a term which underlines the multiple closure properties of the class. A similar construction, with different parameterization, has been put forward by Liseo and Loperfido (2003) in a Bayesian framework, denoted *hierarchical skew-normal*. Another similar distribution is the *fundamental skew-normal distribution* of Arellano-Valle and Genton (2005).

The connections among these constructions have been examined by Arellano-Valle and Azzalini (2006), showing their essential equivalence once they are suitably parameterized and redundancies of parameters are removed. This explains the term *unified skew-normal*, briefly SUN, for their parameterization, which corresponds to (7.2) here. Their Appendix A gives the expressions of the marginal and conditional densities in the parameterization adopted here, that is (7.6) and (7.7). They also provide conditions for independence between blocks of components of Y. The corresponding expression of the mean value is given in an appendix of Azzalini and Bacchieri (2010). The case of the singular SUN distribution, which arises when we drop the assumption that Ω^* is of full rank, has been examined by Arellano-Valle and Azzalini (2006, Appendix C); the resulting distribution can take different forms, depending on which block of Ω^* is the source of the singularity.

The extension of Stein's lemma to multivariate SN variates, presented in Complement 5.3, has been formulated by Adcock (2007) also for closed skew-normal distributions. In their study of two-level hierarchical models subject to a Heckman-type selection mechanism, Grilli and Rampichini (2010) highlight the connection with SN and SUN distributions.

7.1.3 Elliptical parent components

The natural subsequent step is to replace the normality assumption in (7.1) by that of an elliptically contoured distribution. We then apply the argument leading to (7.2) to a $(d+m)$-dimensional elliptical distribution with density generator $\tilde{p}_{d+m}$, say, and arrive at the density

$$p(x) = p_d(x - \xi; \Omega) \, \frac{P_{m, Q(x)}\left\{\tau + \Delta^\top \bar{\Omega}^{-1} \omega^{-1}(x - \xi); \Gamma - \Delta^\top \bar{\Omega}^{-1} \Delta\right\}}{P_m(\tau; \Gamma)} \qquad (7.9)$$

for $x \in \mathbb{R}$, where p_d is the marginal density of the first d components of X, P_m is the marginal distribution function of the remaining m components and $P_{m,Q(x)}$ is the conditional distribution function of the observables given the latent variables, which depends on $Q(x) = (x - \xi)^\top \Omega^{-1}(x - \xi)$, as stated by Proposition 6.1(g). The term *skew unified elliptical distribution* is used referring to (7.9), briefly SUEC by merging SUN and SEC.

An interesting subclass of (7.9) occurs when the parent elliptical distribution arises as a scale mixture of $(d+m)$-dimensional normal variates. Within this subclass, an especially important case arises when the mixing distribution is Gamma$(v/2, /v/2)$, leading to an extension of the multivariate EST distribution, with a density similar to (6.34) but with the $T(\cdot)$ distribution function replaced by its multivariate version $T_m(\cdot)$.

Bibliographic notes

Arellano-Valle and Azzalini (2006) present the formulation of a skew unified elliptical distribution along the lines sketched above, and derive some basic properties, such as closure under marginalization and conditioning. Arellano-Valle and Genton (2010c) have carried out a far more extensive investigation of the formal properties, working with a somewhat different parameterization. Jamalizadeh and Balakrishnan (2010, Section 2) present various results for SUEC$_{1,m}$ distributions, including a form of Student's t-type SUEC.

7.1.4 Some noteworthy special cases

We summarize here a set of distributions which formally are special cases of the SUN, but the root formulation of this group of distributions preceded those of § 7.1.2.

Balakrishnan (2002) has examined univariate densities of the mathematically simple form

$$c_m(\lambda)^{-1} \, \varphi(x) \, \{\Phi(\lambda x)\}^m, \qquad x \in \mathbb{R}, \tag{7.10}$$

where $m \in \mathbb{N}$ and $c_m(\lambda)$ is a normalizing constant which in general depends on the real parameter λ, apart from $c_0 = 1$ and $c_1 = 1/2$. The author derives a recursive relationship for the sequence $c_m(\lambda)$, and specifically obtains

$$c_2(\lambda) = \frac{1}{\pi} \arctan \sqrt{1 + 2\lambda^2}, \quad c_3(\lambda) = \frac{3}{2\pi} \arctan \sqrt{1 + 2\lambda^2} - \frac{1}{4}.$$

Density (7.10) can be viewed as an instance of (7.2) where $d = 1$ and the Φ_m term in the numerator of the fraction represents the distribution

function of m independent $N(0, 1)$ components. More specifically, we set
$\xi = 0$, $\omega = 1$, $\bar{\Omega} = 1$, $\tau = 0$, $\Delta = \lambda 1_m^\top$, $\Gamma = I_m + \Delta^\top \Delta$, so that we
can identify $c_m(\lambda) = \Phi_m(0; I_m + \lambda^2 1_m 1_m^\top)$. Also, Balakrishnan points out a
close connection between $c_m(1)$ and the expected value of the largest-order
statistic in a sample from $N(0, 1)$.

Additional work on distribution (7.10) has been done by Gupta and
Gupta (2004), who have derived lower-order moments and other formal
properties for $m = 2$ and $m = 3$.

Jamalizadeh and Balakrishnan (2008; 2009) replace the final factor in
(7.10) by a standard bivariate normal distribution function $\Phi_B(x, y; \rho)$, arriving at the density

$$c(\lambda_1, \lambda_2, \rho)^{-1} \varphi(x) \Phi_B(\lambda_1 x, \lambda_2 x; \rho), \qquad x \in \mathbb{R}, \qquad (7.11)$$

where the normalizing constant is

$$c(\lambda_1, \lambda_2, \rho) = (2\pi)^{-1} \arccos\left[-\{(1 + \lambda_1^2)(1 + \lambda_2^2)\}^{-1/2}(\rho + \lambda_1 \lambda_2)\right].$$

From the moment generating function, the authors obtain the lower-order
moments and from here, with a special version of Stein's lemma, an expression for marginal moments of any order and other formal properties.
Density (7.11) is of $\mathrm{SUN}_{1,2}$ type with $\tau = 0$, $\omega = 1$ and $\Gamma = R + \Delta^\top \Delta$,
where $\Delta = (\lambda_1, \lambda_2)$ and $\mathrm{vech}(R) = (1, \rho, 1)$. Furthermore, if a variable Z has
distribution (7.11), then $Z/\sqrt{V}$ where $V \sim \chi_\nu^2/\nu$ lends a form of generalized
bivariate skew-t distribution belonging to the $\mathrm{SUEC}_{1,2}$ class.

Sharafi and Behboodian (2008) examine additional properties of (7.10),
including two forms of stochastic representation and a recurrence relationship of moments. In addition they propose a bivariate extension of the Balakrishnan distribution, that is

$$c_m(\lambda, \rho)^{-1} \varphi_B(x_1, x_2; \rho) \{\Phi(\lambda^\top x)\}^m, \qquad x = (x_1, x_2) \in \mathbb{R}^2,$$

where $\lambda = (\lambda_1, \lambda_2)^\top$ and $\varphi_B(x_1, x_2; \rho)$ denotes the bivariate normal density
with standardized marginals and correlation ρ. The correspondence with a
$\mathrm{SUN}_{2,m}$ distribution can be established setting $\omega = I_2$, $\mathrm{vech}(\bar{\Omega}) = (1, \rho, 1)^\top$,
$\Delta = \Delta_0 1_m^\top$ where $\Delta_0 = \bar{\Omega}\lambda = (\lambda_1 + \rho\lambda_2, \lambda_2 + \rho\lambda_1)^\top$, and the other terms
similarly to the case (7.10), in particular

$$\Gamma = I_m + \Delta^\top \bar{\Omega}^{-1} \Delta = I_m + Q(\lambda) 1_m 1_m^\top$$

where $Q(\lambda) = \lambda_1^2 + 2\rho\lambda_1\lambda_2 + \lambda_2^2$. From here it follows that $c_m(\lambda, \rho) = c_m(Q(\lambda)^{1/2})$, where $c_m(\cdot)$ is as in (7.10), as noted by the authors. The
extension to $d > 2$ is immediate.

7.1.5 Connections with order statistics

The distributions examined in the previous chapters, when the auxiliary location parameter τ is zero, can be related to the distribution of the minimum or the maximum of a random sample from an appropriate symmetric distribution, both in the univariate and in the multivariate context. In some cases, additional representations via order statistics hold; cf. Complement 2.3

This type of connection exists also for several distributions described above, notably those summarized in § 7.1.4. For instance, Jamalizadeh and Balakrishnan (2009) show that a stochastic representation for (7.11), in addition to the general ones for the SUN family, is $W_1| \min(W_2, W_3)$ where (W_1, W_2, W_3) is a trivariate normal variable whose correlation matrix is suitably related to the parameters of (7.11). Similar representations exist for several other special cases of the SUN distributions. What is not available at the time of writing is a general representation based on maxima or minima for all SUN distributions (7.2) with $\tau = 0$.

A connection with order statistics arises also from another direction, that is, in the study of the distribution of ordered values of a vector sampled from an elliptical distribution, a problem which can take a variety of different forms. We do not attempt a full discussion of the theme and restrict ourselves to recall two papers, referring the interested reader to these sources for earlier references on the same problem. Both Arellano-Valle and Genton (2007, Proposition 2) and Jamalizadeh and Balakrishnan (2010, Theorem 9) have considered the joint distribution of a linear combination $L Z_{(n)}$ where L is a $p \times n$ matrix of constants and $Z_{(n)}$ is the vector of ordered values of an n-dimensional variable with elliptical distribution of type (6.2). The distribution of $L Z_{(n)}$ turns out to be a mixture of $n!$ components of $\mathrm{SUEC}_{1,p}$ type. It must, however, be noticed that the expressions of the resulting density provided in these sources are not exactly coincident.

7.2 Flexible and semi-parametric formulation

The next two themes are technically slightly different but they appear connected by the common attitude of handling the symmetry-perturbing function with 'high flexibility'.

7.2.1 Flexible skew-symmetric distributions

It is well known that a sufficiently regular function can be closely approximated by a polynomial of adequately high degree. Ma and Genton

(2004) examine a similar problem for the modulation factor of a skew-symmetric distribution of type (1.3). On writing $G(x)$ in the form $G_0\{w(x)\}$ for some given choice of G_0, consider $w(x)$ which is a polynomial such that $w(-x) = -w(x)$. Therefore, consider the family

$$f(x) = 2 f_0(x) G_0\{w_K(x)\}, \qquad x \in \mathbb{R}^d, \qquad (7.12)$$

where $w_K(x)$ is an odd multivariate polynomial in $\mathbb{R}^d$, that is a polynomial with only terms of odd order up to a maximal order K. This family represents a generalization of the FGSN family of § 2.4.3 which referred to the univariate case when the base density is normal and $G_0 = \Phi$. Ma and Genton (2004) present a variety of numerical examples in the univariate and in the bivariate case, using the normal and t as the base density. Selection of the order K is accomplished by an information criterion, such as AIC or BIC.

An interesting question is then: if we let $K \to \infty$, how wide can this family be? The following result by Ma and Genton (2004) shows that (7.12) can be arbitrarily close to any skew-symmetric density.

Proposition 7.2 *Consider densities of the form (7.12) where f_0 and G_0 satisfy the conditions of Proposition 1.1, and $w_K(x)$ is an odd polynomial of order K. The set of densities (7.12) is dense, in the L^∞ norm, in the set of densities (1.3) where $G(x)$ satisfies (1.4) and it is continuous.*

This statement draws a connecting line between the parametric formulation of the perturbation function in (7.12), as K increases, and the general 'non-parametric' form $G(x)$ in (1.3), if they employ the same base function $f_0(x)$.

Instead of odd polynomials as in $w_K(x)$, Genton (2005, Section 8.1) has suggested considering other sets of functions which form an orthogonal basis. He has specifically sketched the use of Fourier sine series of the form

$$w_F(x) = \sum_{m=1}^{M} b_m \sin(m x)$$

and their extensions to the multivariate case.

In a similar logic, Frederic (2011) has suitably adapted the notion of B-splines to the present context. For a set of knots placed symmetrically around 0, two sets of splines are formed, $B^+(x)$ and $B^-(x)$, each with M components, such that $B^+(x) = B^-(-x)$. Then

$$w_B(x) = \left(B^-(x) - B^+(x)\right)^{\mathsf{T}} \beta, \qquad \beta \in \mathbb{R}^M$$

is an odd spline. A multivariate version can be obtained by considering a d-dimensional basis formed by the tensor product of d univariate odd B-spline bases.

In the case of univariate observations $y_1, \ldots, y_n$, the penalized log-likelihood function for location ξ, scale ω, shape v and spline parameters β is

$$\ell(\xi, \omega, v, \beta) = \text{constant} - n \log \omega + \sum_i f_0(z_i; v) + \sum_i \log G_0(w_B(z_i)) - \lambda P(\beta),$$

where $z_i = \omega^{-1}(y_i - \xi)$ and the last term penalizes the 'roughness' $P(\beta)$ of $w_B(\cdot)$ multiplied by a smoothing parameter $\lambda \geq 0$, in the same form as commonly in use in the context of smoothing splines.

7.2.2 Semi-parametric estimation

Consider the case where it is known that the data are not sampled from the distribution of interest f_0 but from a perturbed version of f_0, because of some interfering sample selection mechanism. To simplify the discussion, focus on the case where f_0 is the $N(\mu, \sigma^2)$ density and the observed distribution is of type (1.3), which in this case, after a shift by an amount μ, becomes

$$f(x) = 2\,\sigma^{-1}\varphi(z)\,G(z), \qquad z = \sigma^{-1}(x - \mu) \in \mathbb{R}, \qquad (7.13)$$

where the perturbing factor $G(\cdot)$ satisfies (1.4) but is otherwise unspecified, except at most some regularity conditions. The target of making inference on (μ, σ) without specification of G is quite ambitious – even hazardous, one might think.

Ma *et al.* (2005) have tackled this problem via the theory of regular asymptotically linear (RAL) estimators, assuming that a simple random sample $(y_1, \ldots, y_n)$ from $f(\cdot)$ is available. In this context, it turns out that a RAL estimator corresponds to an even function $t(z) = (t_1(z), t_2(z))$ such that $\mathbb{E}\{t(Z)\} = 0$ if $Z \sim N(0, 1)$ and the estimates are obtained by solving the equations

$$\frac{1}{n} \sum_i t_k \left(\frac{y_i - \mu}{\sigma} \right) = 0, \qquad k = 1, 2, \qquad (7.14)$$

a conclusion which is not surprising, recalling the modulation invariance property (1.12). Substantial work of Ma *et al.* (2005) is dedicated to the choice of the asymptotically optimal function $t(\cdot)$, which is possible if one can posit a specific G.

Essentially the same problem has been considered by Azzalini *et al.* (2010), but with some differences. One is that the argument for considering estimating equations of type (7.14) is taken directly by the modulation-invariance property, which explains their term 'invariance-based estimating equation'. A reasonable and simple choice for $(t_1(z), t_2(z))$ is provided by

$$t_k(z) = |z|^k - c_k, \qquad c_k = \mathbb{E}\{|Z|^k\} = 2^{k/2}\Gamma[(k+1)/2]/\Gamma(1/2) \qquad (7.15)$$

for $k = 1, 2$. One convenient aspect of this option is that this $t_2(z)$ leads to a standard expression for the estimate of σ^2, so that, after substitution of this estimate in the t_1 equation, we must effectively solve a single non-linear equation for μ. While other choices for $t(z)$ are possible, it appears that the crucial point is not so much the choice of $t(z)$, rather the selection of the 'right' root of (7.14), since its roots typically occur in pairs. For each solution of (7.14), $(\hat\mu_j, \hat\sigma_j)$, we can compute a non-parametric estimate $\tilde f_j(z)$ from the normalized residuals $z_{ij} = (y_i - \hat\mu_j)/\hat\sigma_j$ for $i = 1, \ldots, n$ and from here obtain

$$r_j(z) = \frac{\tilde f_j(z)}{2\,\varphi(z)}, \qquad \hat G_j(z) = \frac{r_j(z)}{r_j(z) + r_j(-z)} \qquad (7.16)$$

such that $\hat G_j(z)$ satisfies (1.4), for $j = 1, 2$.

To illustrate the working of the method, reconsider the Barolo phenols data used for Figure 3.1. Solution of (7.14) for $t = (t_1, t_2)$ given by (7.15) produces two pairs of $(\hat\mu_j, \hat\sigma_j)$ estimates: $(2.434, 0.528)$ and $(2.996, 0.371)$. For each of these pairs, estimates $\hat G_j$ have been computed from (7.16) and multiplication by the $N(\hat\mu_j, \hat\sigma_j^2)$ density leads to the continuous function in the top-left and bottom-left panels of Figure 7.1, respectively; the dashed curves are the same non-parametric estimate of Figure 3.1.

To choose one of the outcomes, we can take into consideration the $\hat G_j$ curves, plotted in the right-hand panels of Figure 7.1. The top curve, $\hat G_1$, is associated with a selection mechanism of normal density which seems, on general grounds, more plausible than the other. Also, $\hat G_1$ is less 'complex' than $\hat G_2$ if one considers as a quantifier of the complexity the integral $\int [\hat G_j''(z)]^2 \, dz$, which is far smaller for $\hat G_1$. Finally, notice that the estimates $(2.434, 0.528)$ are much the same as the first two components of $\hat\theta^{DP}$ obtained in § 3.1.2 under SN assumption, and the corresponding $\hat G_1(z)$ resembles a normal distribution function. In some other instances, the difference between the competing estimates of the parameters can be appreciably wider than in this example.

Extensions of this methodology to regression models, skew-t distributions and the multivariate case have been examined. However, the above

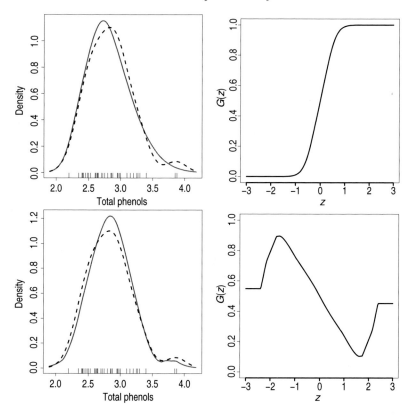

Figure 7.1 Data on total phenols content in Barolo: the left-hand plots display semi-parametric estimates of the density (solid curves) compared with a non-parametric estimate (dashed curves), the right-hand plots display the perturbing function $\hat{G}$.

discussion is intended to stress that, in this setting, special caution should be exercised, and an interplay with subject-matter consideration must be considered whenever possible. Using these tools like an automaton might put into effect the worse implications of the term 'hazardous' employed at the beginning of this section.

Potgieter and Genton (2013) consider the same estimation problem via a different methodology targeted to find the minimum distance between the empirical chracteristic function of the data and the characteristic function of a chosen member of the family (7.13). Although formulated differently, this approach faces the same problem of selecting the 'right' root of the associated equations, much as with equations (7.14).

7.3 Non-Euclidean spaces

7.3.1 Circular distributions

Circular data arise when observations represent a direction, that is an angle, θ, on the unit circle. A classical instance of this type of data is the wind direction recorded at a given geographical location on repeated occasions. Since angles are measured from an arbitrary origin, this calls for specially developed methods, which must neutralize this arbitrariness. In the continuous case, a probability model for this type of data is represented by a circular distribution, $f(\theta)$, which must be a non-negative periodic function, such that $f(\theta+2\pi) = f(\theta)$ for all θ and its integral over an interval of length 2π must be 1. A standard account for the treatment of circular and, more generally, directional data is the book of Mardia and Jupp (1999).

Classical circular distributions are symmetric about a certain angle, which we can take to be 0, without loss of generality. In recent years, more interest has been directed towards asymmetric distributions. Since a traditional circular distribution is the so-called wrapped normal, obtained by 'wrapping' the standard normal distribution around the unit circle, a natural asymmetric analogue replaces the normal density by the skew-normal (2.3), leading to the *wrapped skew-normal* density

$$f_{\text{WSN}}(\theta; \xi, \omega, \alpha) = \sum_{k=-\infty}^{\infty} \frac{1}{\omega}\, \varphi\left(\frac{\theta + 2\pi k - \xi}{\omega}; \alpha\right) \qquad (7.17)$$

proposed by Pewsey (2000b); see Pewsey (2003) and Pewsey (2006a) for additional work. The left plot of Figure 7.2 displays the shape of this distribution for $\alpha = 0, 3, 10$; when $\alpha = 0$ we obtain the classical wrapped normal distribution.

Hernández-Sánchez and Scarpa (2012) consider a similar wrapping construction where the summands correspond to a FGSN distribution, discussed in § 2.4.3, with $K = 3$. There is then an additional shape parameter, which allows us to accommodate the observed bimodal distribution in their applied problem.

For a general formulation more directly connected to (1.2), Umbach and Jammalamadaka (2009; 2010) start from two circular distributions, $f_0(\theta)$ and $g_0(\theta)$, say, both symmetric about 0 so that $f_0(\theta) = f_0(-\theta)$ and $g_0(\theta) = g_0(-\theta)$, and let $G_0(\theta) = \int_{-\pi}^{\theta} g_0(\omega)\, d\omega$. Then they prove that

$$f_{\text{css}}(\theta) = 2 f_0(\theta)\, G_0\{w(\theta)\} \qquad (7.18)$$

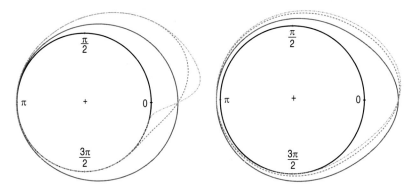

Figure 7.2 Circular distributions: in the left panel wrapped SN with $\alpha = 0$ (solid line), $\alpha = 3$ (dashed), $\alpha = 10$ (dot-dashed); in the right panel a sine-skewed Cauchy distribution with $\lambda = 0, 0.7, 0.99$.

is a circular distribution if w satisfies

$$w(\theta) = -w(-\theta) = w(\theta + 2k\pi) \in [-\pi, \pi)$$

for all θ and all integers k. In plain words, the requirements are those of Proposition 1.1 for the univariate case plus the conditions of periodicity of the component functions and $|w(\theta)| \leq \pi$. If we want to introduce a location parameter, ξ, we replace θ in the right-hand side of (7.18) by $\theta - \xi$.

A modulation invariance property analogous to Proposition 1.4 holds in the following sense: if $h(\theta)$ is a periodic even function with period 2π, then

$$\int_{-\pi}^{\pi} h(\theta) f_0(\theta) \, d\theta = \int_{-\pi}^{\pi} h(\theta) f_{\text{css}}(\theta) \, d\theta. \qquad (7.19)$$

For a circular distribution $f(\theta)$, an important role is played by the trigonometric moments, which are defined as

$$\alpha_p = \int_{-\pi}^{\pi} \cos(p\theta) f(\theta) \, d\theta, \qquad \beta_p = \int_{-\pi}^{\pi} \sin(p\theta) f(\theta) \, d\theta$$

for $p = 0, \pm 1, \pm 2, \dots$ These sequences satisfy $\alpha_{-p} = \alpha_p$ and $\beta_{-p} = -\beta_p$.

Because of (7.19), $f_0(\theta)$ and $f_{\text{css}}(\theta)$ have the same sequence of coefficients α_p. If $w(\theta)$ takes the form $\lambda \bar{w}(\theta)$ where $\lambda \in [-1, 1]$ and $\bar{w}(\theta) \geq 0$ for $\theta \in [0, \pi)$, one can prove that β_1 is an increasing function of λ. This implies that the mean direction increases with λ and the circular variance decreases, similar to distributions on the real line, as discussed in § 1.2.3.

An interesting subset of (7.18) is obtained when G_0 corresponds to the uniform distribution and $\bar{w}(\theta) = \pi \sin \theta$, leading to

$$f_{\text{ssc}}(\theta) = f_0(\theta) \{1 + \lambda \sin(\theta)\} . \tag{7.20}$$

This class of distributions has been examined in the above-mentioned work of Umbach and Jammalamadaka and studied further by Abe and Pewsey (2011) under the heading *sine-skewed circular* distributions. One of their findings is that $\beta_p - \lambda(\alpha_{0,p-1} - \alpha_{0,p+1})/2$, where $\alpha_{0,p}$ is the pth cosine moment of f_0. This allows simple computation of β_p in a range of cases, since explicit expressions of trigonometric moments exist for various classical symmetric distributions. The right panel of Figure 7.2 displays (7.20) when f_0 is the wrapped Cauchy distribution, that is

$$f_0(\theta) = \frac{1}{2\pi} \frac{1 - \rho^2}{1 - 2\rho \cos\theta + \rho^2}$$

with concentration parameter $\rho = 0.5$, and λ in (7.20) takes values $\lambda = 0, 0.7, 0.99$. The resulting sine-skewed distributions exhibit a less visible departure from symmetry than the curves in the left panel of the figure.

7.3.2 Distributions on the simplex

Compositional data arise when we record the proportions of constituents of certain specified types to form a whole. A typical example is represented by the geochemical composition of rocks or other material, such as the proportions of sand, silt and clay in sediments, but clearly there is an enormous range of sources for data of this type. If there are D different types of constituents, each observed unit produces a D-part composition represented by the proportions $p = (p_1, p_2, \ldots, p_D)$ such that

$$p_1 > 0, \ldots, p_D > 0, \qquad \sum_{i=1}^{D} p_i = 1 \tag{7.21}$$

where strict inequalities are indicated, instead of the more general $p_j \geq 0$, in the light of what follows. The sum constraint implies that p is essentially a d-dimensional entity with $d = D - 1$. The geometrical object formed by all points satisfying conditions (7.21) is called the standard d-simplex in $\mathbb{R}^D$, denoted $\mathbb{S}^d$.

The constrained nature of the sample space (7.21) inevitably induces peculiar features for data of this type and it calls for a specifically developed methodology. The standard account for the analysis of compositional data is the monograph of Aitchison (1986). For the purpose of data fitting,

Aitchison proposed mapping the simplex $\mathbb{S}^d$ to $\mathbb{R}^d$ via a suitable invertible transformation, $y = T(p)$ say, and then fitting the transformed data by a d-dimensional normal distribution. A simple and important option for the choice of $T(\cdot)$ is represented by the *additive log-ratio transformation*

$$y_j = \log(p_j/p_D) \qquad (j = 1, \ldots, d), \qquad (7.22)$$

written compactly as $y = \mathrm{alr}(p)$. When a multinormal distribution is assigned to y, this induces a distribution for p on $\mathbb{S}^d$, called additive logistic normal (ALN).

An appeal of this distribution derives from the simple handling of various operations which can be performed on proportions. For instance, one such operation, called sub-composition, consists of extracting from p a subset of $D' < D$ components followed by renormalization to ensure that the new vector p' belongs to $\mathbb{S}^{d'}$, where $d' = D' - 1$. This marginalization process can be expressed nicely, since the resulting distribution is still ALN with parameters which are simple transformations of the original ones.

Mateu-Figueras *et al.* (2005) have extended this construction by replacing the Gaussian assumption with skew-normality. Specifically, if $\mathrm{alr}(p) \sim SN_d(\xi, \Omega, \alpha)$, the induced density for p at $x = (x_1, x_2, \ldots, x_D) \in \mathbb{S}^d$ is

$$f_p(x) = 2\,\varphi_d\big(\mathrm{alr}(x) - \xi; \Omega\big)\,\Phi\left(\alpha^\top \omega^{-1}\{\mathrm{alr}(x) - \xi\}\right)(x_1\,x_2 \cdots x_D)^{-1} \quad (7.23)$$

where the last factor is the Jacobian of the transformation. The authors denote this distribution *additive logistic skew-normal*, which obviously reduces to the ALN when $\alpha = 0$.

A number of the above-mentioned properties of the ALN distribution, such as closure with respect to sub-compositions, carry on for (7.23). These facts are established by suitably exploiting the closure property of the SN class with respect to affine transformations.

Further work using a different type of mapping between $\mathbb{R}^d$ and $\mathbb{S}^d$ in place of (7.22) has been considered briefly by Mateu-Figueras *et al.* (2005) and more extensively by Mateu-Figueras and Pawlowsky-Glahn (2007), where it is advocated to express densities with respect to an alternative measure, more suitable for the simplex, in place of the usual Lebesgue measure.

7.4 Miscellanea

7.4.1 Matrix-variate distributions

In some cases, the outcome of a set of observations is naturally arranged in a matrix, with dependence among observations being associated both with row and column variation. As a typical situation, start by considering a multivariate distribution, which is used to describe the set of p observations taken on a given individual; then consider the case where the same set of variables are recorded from that individual at q occasions along time.

A (p,q)-dimensional variable X is said to have matrix-variate normal distribution, with 0 location, if its density function at $x \in \mathbb{R}^{p \times q}$ is

$$\varphi_{p,q}(x; \Psi \otimes \Sigma) = \frac{1}{(2\pi)^{pq/2} \det(\Sigma)^{q/2} \det(\Psi)^{p/2}} \exp\left[-\frac{1}{2} \operatorname{tr}\left(\Psi^{-1} x^{\top} \Sigma^{-1} x\right)\right] \tag{7.24}$$

where Σ and Ψ are symmetric positive-definite matrices of order p and q, respectively. An additional location parameter takes the form of a $p \times q$ matrix M and is notionally associated with $X' = X + M$ whose density is (7.24) evaluated at $x - M$. From the properties $\operatorname{tr}(AB) = \operatorname{vec}(A^{\top})^{\top}\operatorname{vec}(B)$ and $\operatorname{vec}(ABC) = (C^{\top} \otimes A)\operatorname{vec}(B)$, we obtain the equivalence between the fact that X has distribution (7.24) and $\operatorname{vec}(X) \sim N_{pq}(0, \Psi \otimes \Sigma)$.

From what has just been remarked, Chen and Gupta (2005) observe that a direct extension of the d-dimensional SN density (5.3) to a (d,m)-variate version can be formulated as

$$2\,\varphi_{d,m}(x; \Psi \otimes \Omega)\,\Phi\{\operatorname{tr}(H^{\top}x)\}, \qquad x \in \mathbb{R}^{d \times m}, \tag{7.25}$$

where Ω is as in (5.3), H is a $d \times m$ matrix and now Ψ is $m \times m$. Given the connection with (5.3), a number of similar properties hold for (7.25).

However, for the reasons which have led to the SUN density (7.2), we may want to consider a more general form of the perturbation factor. Harrar and Gupta (2008) propose the density function

$$\frac{1}{\Phi_m(0; \Upsilon + \eta^{\top}\Omega\eta\,\Psi)}\,\varphi_{d,m}(x; \Psi \otimes \Omega)\,\Phi_m(x^{\top}\eta; \Upsilon), \qquad x \in \mathbb{R}^{d \times m}, \tag{7.26}$$

where η is a vector in $\mathbb{R}^m$ and Υ is an $m \times m$ symmetric positive-definite matrix. The authors show that the density is properly normalized, and derive the expression of the moment generating function and a range of formal properties.

7.4.2 Non-elliptical base distribution

So far, when we have dealt with multivariate distributions, the base density f_0 has been in nearly all cases of elliptical type; an exception occurs in Problem 1.5. Arellano-Valle and Richter (2012) have developed an extensive construction where the Euclidean norm, which is at the core of the elliptical class, is replaced by the general L^p-norm, hence working with a non-elliptical base density.

To begin with, recall the Subbotin density (4.1) on the real line, except that here we denote its tail-weight parameter by p. If the components of $X = (X_1, \ldots, X_d)^\top$ are independent and identically distributed Subbotin's variables, the density function of X at $x = (x_1, \ldots, x_d)^\top \in \mathbb{R}^d$ is

$$f_0(x) = c_p^d \exp\left(-\frac{\|x\|_p^p}{p}\right), \qquad (7.27)$$

where

$$\|x\|_p = \left(\sum_{j=1}^d |x_j|^p\right)^{1/p}$$

is the L^p-norm of x. Density (7.27) is a form of multivariate Subbotin distribution, but different from that mentioned near the end of § 6.1.1.

Similarly to spherical variables, one can formulate a representation of type $X = R U_p$, where R is a univariate positive variable and U_p is an independent variable uniformly distributed on the p-generalized unit sphere, that is $\{x : x \in \mathbb{R}^d, \|x\|_p = 1\}$.

The analogy with spherical distributions is taken further by Arellano-Valle and Richter (2012), who extend (7.27) to the case of a more general form of density generator, under a condition similar to (6.1) with r^2 replaced by r^p. From here, they develop a whole construction which generalizes that of spherical distributions, where the L^p-norm replaces the Euclidean norm. If X now has one such distribution, a linear transformation $Y = \mu + \Gamma X$ produces a distribution with constant density on p-generalized ellipsoids.

If Y is partitioned into two blocks and we replicate the conditioning mechanism of § 6.1.2, this lends an extension of the skew-elliptical class (6.15) and correspondingly a set of analogues of its sub-families. An instance of these sub-families is a skew version of the density (7.27), multiplied by the modulation factor given by an m-dimensional distribution function of the same type. This is a multivariate extension of the asymmetric (type I) Subbotin distribution of Complement 4.1.

7.4.3 Bimodal skew-symmetric distributions

Consider functions f_0, G_0 and w satisfying conditions of Proposition 1.1 for $d = 1$, with the additional assumption that $\kappa = \int_{-\infty}^{\infty} x^2 f_0(x)\,dx < \infty$. Since $(1 + \psi x^2) f_0(x)/(1 + \psi\kappa)$ is a symmetric density if $\psi > 0$, Elal-Olivero *et al.* (2009) conclude that

$$f(x) = 2 \, \frac{1 + \psi \, x^2}{1 + \psi \, \kappa} \, f_0(x) \, G_0\{w(x)\}$$

is a proper density function for any choice of $\psi \geq 0$. Alternatively, one can reach the same conclusion using Proposition 1.4 with $t(x) = x^2$. An especially simple case of $f(x)$ occurs when $f_0(x) = \varphi(x)$, so that $\kappa = 1$, and $G_0(x) = \Phi(x)$.

An interesting feature of $f(x)$ is that it can produce both unimodal and bimodal behaviour, depending on the value of ψ, even using a linear form $w(x) = \alpha \, x$. Suppose that, for some observed data, the empirical distribution exhibits bimodal shape or possibly unimodal but with a 'hump'. With the addition of a location and scale parameter, this distribution represents a four-parameter competitor both of the mixture of two normal densities, which involves five parameters, and of the four-parameter FGSN distribution with $K = 3$.

8

Application-oriented work

8.1 Mathematical tools

8.1.1 Approximating probability distributions

Since the SN family embeds the normal distribution, it is quite natural to employ it as an approximating distribution in place of the normal one in a range of cases where the normal approximation is known to work asymptotically, when some index n diverges, but it is not fully satisfactory for finite n.

The simplest example of this sort is provided by the normal approximation to the binomial distribution. To fix notation, consider a variable Y_n having binomial distribution with index n and probability parameter p. For fixed p and diverging n, a standard approximation to the probability distribution of Y_n is provided by a normal distribution with the same first two moments, that is $N(np, np(1 - p))$. For any fixed n, this approximation works best when $p = \frac{1}{2}$, and it degrades if p approaches the endpoints of the interval $(0, 1)$. Since this deterioration of the approximation is related to the asymmetry of Y_n when $p \neq \frac{1}{2}$, one expects that a better outcome will be obtained by adopting an asymmetric distribution as an approximant.

This idea has been examined by Chang *et al.* (2008), who have approximated the binomial by an SN distribution by equating their moments up to order three. Moment matching is possible in explicit form, except in some sporadic cases with p very close to 0 or 1 where no solution exists. Although this approximation lacks a theoretical back-up analogous to the de Moivre–Laplace theorem behind approximation via the normal distribution, still it turns out to be very effective, especially when p is not close to $\frac{1}{2}$, as demonstrated numerically by the authors.

A similar type of approximation has been examined by Chang *et al.* (2008) for other discrete distributions, specifically the negative binomial and the hypergeometric distribution.

Obviously, the SN distribution can be used even more naturally for approximating a continuous distribution. For example, in the method proposed by Guolo (2008) for using prospective likelihood methods to analyse retrospective case-control data, a key role is played by an SN approximation to the distribution of a continuous covariate which is not directly observable and requires to be modelled in a flexible way.

A further step in this direction is to employ an ST distribution, provided the problem under consideration allows us to consider four moments.

A more general treatment, valid also in the multivariate case, of the use of the skew-normal as an approximating distribution has been formulated by Gupta and Kollo (2003). They have developed an expansion of Edgeworth type for approximating a density function where the SN density replaces the normal as the leading term. Remarkably, this expansion is not much more complicated than the classical one built on the normal distribution. Owing to the extensive technical aspects, we do not attempt a more detailed description and refer the reader to the paper of Gupta and Kollo; notice that what they denote by α corresponds to our η.

8.1.2 Approximation of functions and non-linear regression

Thanks to their high flexibility, densities of type (1.2) have been shown to be useful also for approximating functions which are not probability distributions, effectively for the purpose of data fitting.

For the problem of approximating a given real-valued function on $\mathbb{R}^d$, a standard approach is via functions of the form

$$f(x) = \sum_{j=1}^{m} w_j f_0(x - x_j; Q_j), \qquad x \in \mathbb{R}^d, \tag{8.1}$$

where $f_0(z; Q)$, called the radial basis function (RBF), depends on z only via $z^\top Q z$, for some positive-definite matrix Q. Additional regulation of $f(x)$ is provided by the 'weights' w_j and the points x_j, for $j = 1, \ldots, m$, suitably chosen to fit the target function. Besides approximating a function, (8.1) can be used to fit observed data in a non-linear regression context, regarding the terms w_j, x_j and Q_j as parameters to be estimated.

Therefore f_0 has much in common with the elliptical distributions of § 6.1.1. In fact, one of the more commonly employed RBFs is the Gaussian, but also other elliptical distributions are in use, although with different names, such as inverse multiquadrics for the scaled Student's t. Therefore (8.1) is similar to a finite mixture of elliptical distributions, but here f_0 does not need to integrate to 1, since the weights are unrestricted, and in general it does not even need to integrate.

Jamshidi and Kirby (2010) have extended (8.1) by replacing f_0 with a 'skewed' version, denoted sRBF, obtained from the constructive tools described in Chapter 1. In the authors' words, 'the statistical literature concerning skew multivariate distributions provides a blueprint for constructing sRBFs'. More specifically, the sRBF notion is formulated in its broader meaning referring to the general form (1.28), but most of the actual development proceeds by replacing $f_0(x - x_j; Q_j)$ in (8.1) with

$$f_0(x - x_j; Q_j) G_0\{\eta_j^\top (x - x_j)\}$$

where η_j is an additional vector parameter. This function is of the linear skew-elliptical type (6.11) after shifting it from 0 to x_j and disregarding the normalization constant, not required in this context because of the factor w_j. Since the intended use is purely data fitting, considerations of numerical efficiency give preference to a choice of G_0 which is simple to compute, such as the Cauchy distribution function. Parameter estimation can be performed by non-linear least squares. Numerical work illustrates the improvement over use of classical RBFs, notably in the reduced number m of terms required.

In a broadly similar logic, but in an independent work and context, Mazzuco and Scarpa (2013) use FGSN distributions of § 2.4.3 for fitting purposes. In this case, the problem is to fit a set of curves which represent age-specific female fertility, typically for a given country along a range of years, and it is desirable that the same type of curve is used as the years progress, varying only its parameters. Fertility curves behave similarly to those of the densities in Figure 2.8, with the qualitative difference that they do not integrate to 1; hence there is an additional multiplicative parameter of the whole curve. More specifically, Mazzuco and Scarpa use curves with base $f_0 = \varphi$, $G_0 = \Phi$ and $K = 3$ in (2.54), which can then be written as

$$w_3(z) = \alpha_t z + \beta_t z^3$$

where α_t and β_t depend on the year t under consideration. After non-linear least-squares fitting, the outcome compares favourably with existing proposals in the literature, for a range of cases referring to different countries.

8.1.3 Evolutionary algorithms

Evolutionary algorithms constitute a set of optimization techniques inspired by the idea of biological evolution of a population of organisms which adapts to their surrounding environment via mechanisms of mutation and selection. An algorithm of this type starts by choosing an initial 'population' formed by a random set of n points in the feasible space of the target function and making it evolve via successive generations. In the evolution process, the points which are best performing are used to breed a new generation via a mutation operator. This step involves generation of new random points, typically making use of a multivariate Gaussian distribution.

In this framework Berlik (2006) has considered adopting a non-symmetric parent distribution, instead of a Gaussian one. The main idea of *directed mutation* is to impart directionality to the search by generating random numbers that lie preferably in the direction where the optimum is presumed to be. Operationally, the SN family provides the sampling distribution, in its univariate or a multivariate version, depending on whether we want to keep mutation in the various components independent or allow for correlation. In the iterative process also the parameters of the n sampling distribution are subject to mutation. For instance, for the univariate SN distribution, the slant parameter of the ith point is modified from α_i to $(1 - k)\alpha_i + z_i$, where $z_i \sim N(0, 1)$ and k is a fixed tuning parameter ($0 \leq k \leq 1$). Berlik (2006, p. 186) reports that directed mutation 'clearly outperformed the other mutation operators'.

8.2 Extending standard statistical methods

In earlier chapters we presented statistical methods for the distributions under consideration in the case of a simple sample or of linear regression, but many other statistical methods have been re-examined in this context.

8.2.1 Mixed effects models

A step beyond the linear regression models considered in earlier chapters is represented by linear mixed models, typically introduced for the analysis of longitudinal data. Given a set of N subjects from which a response variable is recorded at successive time points, the typical form of a linear mixed model for the n_i-vector of responses y_i observed on the ith subject is

$$y_i = X_i\beta + Z_i b_i + \varepsilon_i, \tag{8.2}$$

where X_i and Z_i denote matrices of covariates having dimension $n_i \times p$ and $n_i \times q$, respectively, β is a p-vector parameter, b_i is a q-vector of individual random effects and ε_i is an n_i-vector of error terms, that is with independent components, and ε_i is independent of b_i. Equation (8.2) holds for $i = 1, \ldots, N$ and it is assumed that distinct subjects behave independently.

In the classical formulation, the random terms are assumed to have multivariate normal distribution. In a fairly standard set-up, $b_i \sim N_q(0, \Sigma)$ and $\varepsilon_i \sim N_{n_i}(0, \psi I_{n_i})$ for some matrix $\Sigma > 0$ and some positive ψ. More elaborate versions let these parameters vary with i as functions of some other parameters and additional covariates, or $\mathrm{var}\{\varepsilon_i\}$ may correspond to a time-series structure, but this does not change the essence of the formulation.

Arellano-Valle *et al.* (2005a) have replaced the normality assumption by skew-normality in three possible forms: for b_i or for ε_i or for both of them. In either of the first two cases, $u_i = Z_i b_i + \varepsilon_i$ is the sum of a skew-normal and a normal variate, whose distribution is skew-normal by Proposition 5.4, which also indicates how to compute its parameters. A more complex situation arises in the third case, when both b_i and ε_i are SN with non-null slant parameter, since u_i is not SN any longer. Arellano-Valle *et al.* (2005a) have obtained the distribution of u_i working from first principles, but we can now make use of the formulation in § 7.1.2 and state directly that, since u_i is the sum of two independent SN_{n_i} variates, that is $SUN_{n_i,1}$, its distribution is type $SUN_{n_i,2}$ and its parameters can be computed via (7.8).

In all three cases concerning the distribution of ε_i and b_i, the log-likelihood is computed as the sum of N individual terms, given independence among subjects. Its maximization can be pursued by direct numerical search using numerical methods or via a form of EM algorithm. The latter is the route adopted by Arellano-Valle *et al.* (2005a) for the first two of the above three types of distributional assumptions on the random terms.

Arellano-Valle *et al.* (2007) have considered the same model as above but with a different form of skew-normal distribution, namely (6.43). In this setting, evaluation of the density function of y involve terms of type $\Phi_{n_i}(\cdot)$, whose computation is unfeasible in many practical instances. Hence, while MLE is feasible, as indicated on p. 192, the model is more naturally fitted using MCMC in a Bayesian context.

In a more elaborate formulation, there is a multivariate response variable. For instance, blood circulation of patients may be better described by employing several variables, not just systolic pressure. In these cases, we may introduce a set of equations of type (8.2), one for each component of the response. Ghosh *et al.* (2007) have dealt with the case of a bivariate response, $(y_i^{(1)}, y_i^{(2)})$ say, hence with two simultaneous expressions of type (8.2). Correspondingly, the joint distribution of the two individual random effects, $(b_i^{(1)}, b_i^{(2)})$ say, is now $(2q)$-dimensional. Also in this case the authors adopt a distribution of type (6.43) and, for the same reasons as above, inference is tackled via MCMC, working in a Bayesian framework.

Bolfarine *et al.* (2007) developed influence diagnostics for linear mixed models when the error terms ε_i in (8.2) have multinormal distribution and the random effects b_i have a multivariate SN distribution, which they parameterize in a form similar to (5.30). Further work on this theme has been done by Montenegro *et al.* (2009).

Extensions of the above constructions of the case of skew-elliptical distributions, with special emphasis on the ST, have been considered too. Jara *et al.* (2008) have examined this direction working with (6.42) in a Bayesian context.

In the classical perspective, Zhou and He (2008) have considered a model where b_i in (8.2) is formed by a single column, the random intercept b_i is univariate ST, $Z_i = 1_{n_i}$, and ε_i is formed by a set of values randomly sampled from another univariate ST distribution, with independence between b_i and all ε_i components. Not surprisingly, given the presence of two distinct parameters regulating the tail weight of the random terms, MLE exhibits instability and the authors develop a special three-step estimation procedure. Ho and Lin (2010) retain instead the general form (8.2) and assume that (b_i, ε_i) is ST_{n_i+q} of type (6.23), up to a change of scale. Maximum likelihood estimation takes place via a variant form of the EM algorithm.

The motivating application of Nathoo (2010) arises from forestry, specifically from a 10-year longitudinal study on tree growth in a plantation. In this problem, various features of the data must be accounted for: a spatial effect among nearby trees, a time effect due to the longitudinal nature of the study and markedly non-Gaussian distribution of the distribution of the observed height. The author formulates a spatio-temporal model which accounts for these two components, and remarks that unobservable random components can reasonably be accounted for assuming normality. If ξ_{it} denotes the combined outcome of the fixed and spatio-temporal random effects, the observed height of the tree at location i at time t is modelled as $ST(\xi_{it}, \omega^2, \alpha, \nu)$. The final outcome confirms a long-tailed distribution with a moderate but significant asymmetry of the ST distribution.

8.2.2 Finite mixtures of distributions

A finite mixture of distributions is obtained as a linear combination of distributions with non-negative weights $\pi_1, \ldots, \pi_K$ which represent the probabilities of the component subpopulations; hence $\sum_k \pi_k = 1$ holds. Usually, the component distributions are taken to be members of the same parametric family; in the continuous case below, this family is represented by density f_c. Hence a finite mixture density takes the form

$$f(x; \pi, \theta) = \sum_{k=1}^{K} \pi_k \, f_c(x; \theta_k), \qquad x \in \mathbb{R}^d, \tag{8.3}$$

where θ_k represents the set of parameters of the kth subpopulation, π comprises $\pi_1, \ldots, \pi_K$ and θ comprises $\theta_1, \ldots, \theta_K$. A standard account for finite mixture models is the book of McLachlan and Peel (2000).

The theme of finite mixtures overlaps with that of model-based cluster analysis, where clusters are associated with the subpopulations. After a mixture of type (8.3) has been fitted to a set of data $x_1, \ldots, x_n$, allocation of an observed point x_i to one of the K clusters is made on the basis of the posterior probabilities $\pi_k \, f_c(x_i; \theta_k)/f(x_i; \pi, \theta)$ evaluated with θ_k and π_k equal to their estimated values, for $k = 1, \ldots, K$.

Predictably, the classical and most developed formulation of type (8.3) takes f_c to be the Gaussian distribution. Another distribution in common use is the Student's t. Both these families have contour-level sets which are ellipsoids, however. This constraint may require that, to accommodate the shape of a non-ellipsoidal data cloud, the fitting process needs to allocate two or more components $f_c(x; \theta_k)$ while a more flexible parametric family might achieve the same result with a single component. In this way, we could reduce the number of components and parameters required to achieve the same level of approximation in the description of the data. It is advisable that, although flexible, f_c remains unimodal for all possible choices of θ, since multimodality is accounted for by the presence of multiple components in (8.3).

A number of authors have proposed replacing the Gaussian assumption for f_c with that of an SN or ST distribution, possibly using their variant form (6.42). Earlier work has dealt with the univariate case; see Lin *et al.* (2007a; 2007b), Basso *et al.* (2010). This was soon followed by consideration of the multivariate case; see Pyne *et al.* (2009), Frühwirth-Schnatter and Pyne (2010), Lin (2009; 2010), Cabral *et al.* (2012).

Most of the above publications include numerical illustrations using some real data. The paper of Pyne *et al.* (2009) presents a fully fledged application to flow cytometry data to which the authors fit a finite mixture where f_c is of multivariate ST type, as described in §6.2. After ML estimation, this leads to a clustering of the data which the authors find markedly preferable to that obtained by classical alternatives based on symmetric distributions. This preference is due to both employing a smaller number K of components and at the same time a more accurate fit to the data distribution. The related paper of Frühwirth-Schnatter and Pyne (2010) employs the same type of formulation but adopts a Bayesian inferential paradigm; applications are again to flow cytometry data and in more classical medical statistics.

In the above-quoted contributions, maximum likelihood estimation is systematically tackled via some instance of the EM family of algorithms, while Bayesian analysis is carried out using MCMC, typically via Gibbs sampling. In all formulations, an additive representation of type (5.19) plays a key role, possibly combined with multiplication by a random scale factor in the case of ST distribution.

8.2.3 Time series and spatial processes

Consider the problem of introducing a stationary discrete-time process $\{Y_t\}$ having SN distribution, either marginally or jointly for blocks of type $(Y_t, Y_{t+1}, \ldots, Y_{t+m-1})$ for some given m. The key issues can be illustrated in the case of the simple autoregression

$$Y_t = \rho Y_{t-1} + \varepsilon_t, \qquad t = 0, \pm 1, \pm 2, \ldots, \tag{8.4}$$

where $-1 < \rho < 1$ and ε_t is independent of all Y_s for $s < t$. Assume for a moment that $Y_{t-1} \sim \text{SN}(\xi, \omega^2, \alpha)$ for some non-zero α; without loss of generality, assume $\alpha > 0$. Clearly, ρY_{t-1} is still of the same type, up to a change of the location and scale parameters, but the same α.

If we now take ε_t to have normal distribution, $\rho Y_{t-1} + \varepsilon_t$ will have an SN distribution but with smaller α, by Proposition 2.3. Replicating the argument for Y_{t+1}, this will have an even smaller α, and so on for $Y_{t+2}, Y_{t+3}, \ldots$

Alternatively, take ε_t to have SN distribution. In this case $\rho Y_{t-1} + \varepsilon_t$ is the sum of two independent SN variables, hence of type $\text{SUN}_{1,2}$. At the next iteration of (8.4) we shall get a $\text{SUN}_{1,3}$ distribution, and so on.

Under both assumptions on the sequence $\{\varepsilon_t\}$, the marginal distribution of Y_{t-1} does not reproduce itself after repeated applications of (8.4) and we do not obtain a stationary distribution of SN type. See Pourahmadi (2007) for a detailed discussion of this problem in connection with ARMA models and for related issues.

There exists a form of autoregression which lends an SN marginal distribution, that is the threshold autoregression (2.50) on p. 43. This sort of formulation has been considered further by Tong (1990, pp. 140–146), in particular the multivariate version of (2.50) which leads to a multivariate SN stationary distribution. Although these results are mathematically most elegant, the peculiar form of non-linear autoregression does not seem suitable for common applications.

Alternatively, consider a construction which exploits the additive representation (2.14). Given two independent stationary normal processes $\{W_{0,t}\}$ and $\{W_{1,t}\}$ with $N(0, 1)$ marginal distribution, define

$$Z_t = \sqrt{1 - \delta^2}\, W_{0,t} + \delta\, |W_{1,t}| \tag{8.5}$$

for some fixed δ where $-1 < \delta < 1$. By construction, Z_t has univariate marginal distribution $SN(0, 1, \alpha(\delta))$. In applied work, one includes location and scale parameters in the form

$$Y_t = \xi_t + \omega_t Z_t \tag{8.6}$$

where ξ_t and ω_t may possibly be regulated by time-dependent covariates.

A continuous-time formulation analogous to (8.5) has been examined by Corns and Satchell (2007). In this case, $\{W_{0,t}\}$ and $\{W_{1,t}\}$ are independent Brownian motions, so that their variance is t, and marginally $Z_t \sim SN(0, t, \alpha(\delta))$. It is shown that $\{Z_t\}$ is a form of skew-Brownian motion in the Itô–McKean sense. After inclusion of location and scale parameters, the authors use this formulation to tackle the classical problem in quantitative finance of overcoming inadequacy of the Black–Scholes pricing formula connected to the underlying assumption of Brownian motion, and obtain a more general expression which allows for the presence of skewness.

For the analysis of spatial data, Zhang and El-Shaarawi (2010) have studied a model of type (8.5)–(8.6) where now the subscript t denotes a point of $\mathbb{R}^m$, for $m > 0$. The two components, $\{W_{0,t}\}$ and $\{W_{1,t}\}$, are assumed to be independent Gaussian spatial processes of similar dependence structure but with different parameter values. Estimation can be pursued via an EM algorithm, regarding $\{W_{1,t}\}$ as the 'missing observation'. For the prediction of the process at a nominated point $s \in \mathbb{R}^m$, given the observed values of $\{Y_t\}$ at points $t_1, \ldots, t_n$, the authors obtain the conditional mean $\mathbb{E}\{Y_s | Y_{t_1}, \ldots, Y_{t_n}\}$.

Notice that, while the marginal distribution of Z_t in (8.5) is SN, the joint distribution of $(Z_{t_1}, Z_{t_2}, \ldots, Z_{t_n})$ at 'time' points $t_1, \ldots, t_n$ is not multivariate SN. A jointly SN distribution could be achieved by restraining the first component to be a fixed random variable, $W_{1,t} \equiv W_1$ say, but this choice has the unpleasant side-effect of a non-vanishing correlation at high lags, because of the persistent component W_1.

Kim and Mallick (2004) and Kim *et al.* (2004) have developed a formulation based on the assumption of existence of a stationary spatial process $\{Z_t\}$ such that, at any set of points $t_1, \ldots, t_n$, the joint distribution of $(Z_{t_1}, Z_{t_2}, \ldots, Z_{t_n})$ is multivariate SN, without commitment to a specific type of construction to achieve this distribution.

This formulation has been refuted by Minozzo and Ferracuti (2012), who have shown that the assumption of joint stationary SN distribution at any set of points, $t_1, \ldots, t_n$, is not tenable, because it runs into coherence problems when one marginalizes the joint distribution over a smaller set of points. The authors underline similar problems also with various other proposals in the literature.

In financial applications, specialized models for time series are used, notably the ARCH model and its variants. Since the presence of skewness in financial time series is a feature not easily accounted for by classical formulations, the tools discussed here become natural candidates for consideration.

De Luca and Loperfido (2004) and De Luca *et al.* (2005) have constructed a GARCH formulation for multivariate financial time series where asymmetric relationships exist among a group of stock markets, with one market playing a leading role over the others. Their construction links naturally with the concepts implied by the multivariate SN distribution, when one considers the different effect on the secondary markets induced by 'good news' and 'bad news' from the leading market.

Corns and Satchell (2010) have proposed a GARCH-style model where the random terms have SN distribution, regulated by two equations, one as in usual GARCH models which pertains to the scale factor, conditionally on the past, and an additional equation of analogous structure which regulates the skewness parameter.

A variant form of Kalman filter for closed skew-normal variates has been studied by Naveau *et al.* (2004; 2005). In this formulation a subset of the state variables is used to regenerate the skewing component at each cycle, avoiding the phenomenon of fading skewness which occurs for the simpler construction of §2.2.3.

8.2.4 Miscellanea

There are still many other contributions in applied areas or extensions of existing methods connected to the themes presented here, but it is not possible to provide an adequate summary within the planned extent of this work. However, we would like to mention at least very briefly the existence of developments in other directions.

Statistical quality control represents another area where many classical techniques rely on the assumption of normality, often made for convenience; hence a more flexible and mathematically tractable assumption is of interest. Tsai (2007) has developed control charts for process control under SN assumption of the quality characteristic. In reliability theory, a recurrent theme is the strength–stress model which is connected to $\mathbb{P}\{X < Y\}$ where Y represents the strength of a component subject to a stress X. Gupta and Brown (2001) study this problem when the joint distribution of (X, Y) is bivariate SN with correlated components; Azzalini and Chiogna (2004) work with independent variables, of which one is SN and the other is normal.

The discussion in §3.4.2 has illustrated the natural connection of the SN distribution with stochastic frontier analysis. The distribution theory developed in previous chapters allows us to reconsider that problem in a more general formulation. Domínguez-Molina *et al.* (2004) employ the closed skew-normal in a few variant settings. Tchumtchoua and Dey (2007) work with the variant form of skew-t of Complement 6.4 where it is implied that the production units do not operate independently.

The theme of adaptive designs for clinical trials, in the case of continuous response variables, has a direct connection with our treatment, because it involves consideration of an event on a certain variable, X, observed in the first stage of the study and a correlated response variable, Y, which is examined conditionally on some event $X \in C$. In a much simplified formulation, the conditioning event may be of the form $X_1 - X_2 > 0$ where X_1 and X_2 represent summary statistics of the end-point for the two arms of a phase II study and, depending on whether $X_1 - X_2 > 0$ is true or false, a certain component Y_1 or Y_2 of the end-point of a phase III study, correlated with the X_j's, becomes the variable of interest and in fact the only one available in the second stage. This mechanism is closely linked with the stochastic representation by conditioning, which we have encountered repeatedly. Specifically, under joint normality of the unconditional distributions, Shun *et al.* (2008) developed a 'two-stage winner design' between two competing treatments and this process involves naturally the ESN distribution; Azzalini and Bacchieri (2010) considered a similar problem when several doses or treatments are compared in the first stage, leading to consideration of a SUN distribution.

8.3 Other data types

8.3.1 Binary data and asymmetric probit

Consider independent Bernoulli variables $B_1, \ldots, B_n$, taking value 0 and 1, with probability of success $\pi_i = \mathbb{P}\{B_i = 1\}$ which depends on a p-vector of covariates x_i via $\pi_i = F(\eta_i)$, where $\eta_i = x_i^\top \beta$ and $F(\cdot)$ is some cumulative distribution function. Given observed data $b_1, \ldots, b_n$, the likelihood function for β is

$$L(\beta) = \prod_{i=1}^{n} F(\eta_i)^{b_i} \{1 - F(\eta_i)\}^{1-b_i}, \tag{8.7}$$

once a specific F has been selected. The most common choices for F are the standard logistic distribution function and $F = \Phi$, leading to logistic regression and probit regression, respectively. In this context, F^{-1} represents the link function of the implied generalized linear model.

Chen *et al.* (1999) and Chen (2004) have proposed employing the SN distribution function as F, leading to a form of asymmetric probit link function. A stochastic representation of this formulation starts from independent N(0, 1) variates U_i and U_i', so that $V_i = |U_i'| \sim \chi_1$, and the derived variable

$$W_i = \eta_i + U_i + \alpha V_i = \eta_i + \sqrt{1 + \alpha^2} \left(\sqrt{1 - \delta(\alpha)^2}\, U_i + \delta(\alpha)\, V_i \right) \tag{8.8}$$

where $\alpha \in \mathbb{R}$, and $\delta(\alpha)$ is as in (2.6). From the additive representation (2.14), we see that $W_i \sim \mathrm{SN}(\eta_i, 1 + \alpha^2, \alpha)$. Finally, write

$$B_i = \begin{cases} 1 & \text{if } W_i \geq 0, \\ 0 & \text{otherwise}, \end{cases} \tag{8.9}$$

so that $F(\eta_i)$ in (8.7) is $\Phi(\eta_i / \sqrt{1 + \alpha^2}; -\alpha)$.

From representation (8.9), we can also write the likelihood function, conditional on the values $v_1, \ldots, v_n$ assumed by the variables $V_1, \ldots, V_n$. This is similar to that of a standard probit regression model,

$$L_c(\beta, \alpha | V_i = v_i) = \prod_{i=1}^{n} \Phi(x_i^\top \beta + \alpha v_i)^{b_i} \{1 - \Phi(x_i^\top \beta + \alpha v_i)\}^{1-b_i},$$

for an extended set of covariates x_i, v_i and of parameters β, α. Regarding the v_i's as the missing part of the 'complete data' allows us to formulate an EM algorithm for maximum likelihood estimation. The same expression also provides the basis for constructing a Gibbs sampler in a Bayesian approach.

The introduction of the latent variable W_i offers a simple route for dealing with the case of ordinal response variable instead of dichotomous. If there are K possible levels of the response, this amounts to splitting the real axis into K non-overlapping intervals, introducing $K-1$ threshold values in (8.9) instead of 1.

A related construction has been considered by Bazán *et al.* (2006) in connection with item response analysis. In this context, a test comprising k items is submitted to each individual in a set of n, in order to examine their abilities. In a formulation commonly in use, the probability of a successful outcome of subject i on item j is written as $F(\eta_{ij})$, where $\eta_{ij} = a_j q_i - b_j$ depends on the individual ability q_i and parameters a_j and b_j of the item, denoted discrimination and difficulty, respectively. Bazán *et al.* (2006) introduce a stochastic formulation similar to (8.8), slightly varied to the form

$$W_{ij} = \eta_{ij} - \left(\sqrt{1 - \delta(\alpha_j)^2} U_{ij} + \delta(\alpha_j) V_{ij} \right) \sim \mathrm{SN}(\eta_{ij}, 1, -\alpha_j)$$

in an obvious extension of the earlier notation. The probability of success is now $\mathbb{P}\{W_{ij} \geq 0\} = \Phi(\eta_{i,j}; \alpha_j)$. This stochastic representation is the basis of the Gibbs sampler employed by the authors for Bayesian inference.

Kim (2002) has considered an asymmetric link function of similar type using the ST distribution instead of the SN. A stochastic representation similar to (8.8) holds once a suitable random scale factor is introduced.

Stingo *et al.* (2011) arrived at a formulation similar to those above via a constructive argument which connects with the Heckman selection model recalled in § 3.4.1 and in § 6.2.7. As in (3.40), a latent variable W serves to select the subset of the population on which a component Y is examined. However, at variance with § 3.4.1, in this case we do not observe Y directly but only observe the indicator variable $B = I_{[\mu,\infty)}(Y)$, which says whether Y exceeds the mean value μ of the error term $\sigma \varepsilon_1$ conditionally on $W > 0$. Since the distribution of Y in (3.40) is now of ESN type, the corresponding link function for π_i is the inverse of the ESN distribution function.

8.3.2 Frailty models for survival data

In survival data analysis, Cox's proportional hazards model plays a fundamental role, and it provides the basis for a variety of extensions. One of the more important developments is to incorporate the presence of unobservable random effects. In its basic version, it is assumed that subjects constituting a homogeneous group (or cluster) of subjects share the value taken by some latent variable W, called *frailty*, which influences the survival time in a constant manner within a given group, but differently in separate groups. Correspondingly, we write the hazard function for the survival time T_{ij} of the jth subject in the ith group as

$$h(t_{ij}) = h_0(t_{ij}) \, w_i \, \exp(x_{ij}^\top \beta) = h_0(t_{ij}) \, \exp(b_i + x_{ij}^\top \beta), \qquad (8.10)$$

where w_i is the value taken by W in the ith group, h_0 is the base-line hazard function, x_{ij} is a vector of covariates, β is a p-dimensional parameter of fixed effects and $b_i = \log w_i$. Since the term $x_{ij}^\top \beta$ usually incorporates an intercept term, there must be no free location parameter in the distribution of the log-frailty, $B = \log W$.

A point of interest is the dependence structure of survival times within the same group, and the choice of the frailty distribution is considered crucial to produce correct inferences on the dependence. Since frailties are not observable, the use of a flexible assumption on the distribution of W is considered a safeguard for this problem.

The distributions discussed in the previous chapters provide natural candidates for the frailty model. One such formulation has been developed by Sahu and Dey (2004) who assume that b_i in (8.10) is a value sampled from $B \sim \mathrm{ST}(0, 1+\alpha^2, \alpha, \nu)$, that is, with interconnected scale and slant parameters. As a measure of the dependence structure, they consider the correlation between log survival times. Under the Weibull assumption of the base-line hazard, this correlation equals $\mathrm{var}\{B\}/(\mathrm{var}\{B\} + \pi^2/6)$, which can be computed explicitly using (4.17) if $\nu > 2$. The proposed inferential procedure is set in the Bayesian framework, through the MCMC methodology, and is illustrated with two numerical examples taken from medical statistics.

A similar problem has been studied by Callegaro and Iacobelli (2012) but with various differences in the formulation. One is that the distribution of B is assumed to be SN. The distribution is parameterized so as to have $\mathbb{E}\{B\} = 0$, and with regulating parameters the standard deviation, σ, and α; this set-up is achieved by making use of (2.22) and (2.23). The authors find that this mix of direct and centred parameters, σ and α, is well suited for their purposes. Another point of difference from the earlier formulation is that here the dependence structure is examined via the cross-ratio function, which for the case of two failures is equivalent to

$$CR(t_1, t_2) = h(t_1 | T_2 = t_2)/h(t_1 | T_2 > t_2) \,,$$

dropping the subscript i. When $CR(t, t)$ is plotted against the distribution function, the resulting curve under the SN assumption on B can reproduce, as α varies, the essential behaviour of each curve associated with other distributions in common use, specifically the normal, the positive stable and the log-transformed Gamma distributions. Parameter estimation is performed via a form of EM algorithm and is illustrated with a real-data problem from medical statistics.

> *E quindi uscimmo a riveder le stelle.*
> *(Inferno XXXIV, 139)*

Appendix A

Main symbols and notation

$N(\mu, \sigma^2)$ the univariate normal (Gaussian) distribution with mean value μ and variance σ^2

$N_d(\mu, \Sigma)$ the d-dimensional normal (Gaussian) distribution with mean vector μ and variance matrix Σ

$SN(\xi, \omega^2, \alpha)$ the univariate skew-normal distribution with direct parameters ξ, ω^2, α (*)

$SN_d(\xi, \Omega, \alpha)$ the d-dimensional skew-normal distribution with direct parameters ξ, Ω, α (*)

$ST(\xi, \omega^2, \alpha, \nu)$ the univariate skew-t distribution with direct parameters $\xi, \omega^2, \alpha, \nu$ (*)

$ST_d(\xi, \Omega, \alpha, \nu)$ the d-dimensional skew-t distribution with direct parameters ξ, Ω, α, ν (*)

$EC_d(\xi, \Omega, \tilde{f})$ d-dimensional elliptical(ly contoured) distribution

$SEC_d(\xi, \Omega, \tilde{f})$ d-dimensional skew-elliptical distribution

χ^2_ν the chi-square distribution with ν d.f.

$\varphi(x)$ the $N(0, 1)$ probability density function at x

$\Phi(x)$ the $N(0, 1)$ distribution function at x

$\varphi_B(x, y; \rho)$ the density function of a bivariate normal variate with stardard marginals and correlation ρ, at $(x, y) \in \mathbb{R}^2$

$\varphi_d(x; \Sigma)$ the $N_d(0, \Sigma)$ density function evaluated at $x \in \mathbb{R}^d$

$\Phi_d(x; \Sigma)$ the $N_d(0, \Sigma)$ distribution function evaluated at $x \in \mathbb{R}^d$

$\varphi(x; \alpha)$ the $SN(0, 1, \alpha)$ probability density function at x (*)

$\Phi(x; \alpha)$ the $SN(0, 1, \alpha)$ distribution function at x (*)

$\varphi_d(x; \Omega, \alpha)$ the $SN_d(0, \Omega, \alpha)$ density function at $x \in \mathbb{R}^d$ (*)

$\Phi_d(x; \Omega, \alpha)$ the $SN_d(0, \Omega, \alpha)$ distribution function at $x \in \mathbb{R}^d$ (*)

(*) When an additional parameter is present, the 'extended form' of the distribution is implied.

$t(x; \nu)$	the Student's t density function with ν d.f. at $x \in \mathbb{R}$
$T(x; \nu)$	the Student's t distribution function with ν d.f. at x
$t(x; \alpha, \nu)$	the ST$(0, 1, \alpha, \nu)$ density function at $x \in \mathbb{R}$ (*)
$t_d(x; \Omega, \alpha, \nu)$	the ST$_d(0, \Omega, \alpha, \nu)$ density function at $x \in \mathbb{R}^d$ (*)
$I_A(x)$	the indicator function of set A
$\Gamma(x)$	the Gamma function
$\zeta_k(x)$	the kth derivative of $\log\{2\,\Phi(x)\}$, see p. 30
$\mathbb{P}\{\cdot\}$	probability
$\mathbb{E}\{\cdot\}$	expected value
var$\{\cdot\}$	variance, variance matrix
cov$\{\cdot\}$	covariance
cor$\{\cdot\}$	correlation, correlation matrix
$\overset{\mathrm{d}}{=}$	equality in distribution
det(A)	determinant of matrix A
$A^\top$	transpose of matrix A
A^{-1}	inverse of matrix A
I_n	the identity matrix of order n
1_n	the $n \times 1$ vector with all 1's
vec(A)	the vector formed by stacking the columns of A
vech(A)	the vector formed by stacking the lower triangle, including the diagonal, of a symmetric matrix A
$\otimes$	the Kronecker product of matrices
$\odot$	the entry-wise or Hadamard product of matrices
$\bar{\Sigma}, \bar{\Omega}, \ldots$	correlation matrices of variance matrices $\Sigma, \Omega, \ldots$
$L(\theta), L(\theta; y)$	the likelihood (function) of θ when y has been observed
$\ell(\theta), \ell(\theta; y)$	the log-likelihood of θ when y has been observed
MLE	maximum likelihood estimate/estimation
$\mathcal{J}(\hat{\theta})$	observed information matrix
$\mathcal{I}(\theta)$	expected information matrix
θ^{DP}	direct parameters
θ^{CP}	centred parameters

(*) When an additional parameter is present, the 'extended form' of the distribution is implied.

Appendix B

Complements on the normal distribution

The univariate normal distribution

A continuous random variable X with support on the real line is said to have a standard normal, or Gaussian, probability distribution if its density function is

$$\varphi(x) = \frac{1}{\sqrt{2\pi}} \exp\left(-\tfrac{1}{2}x^2\right), \qquad -\infty < x < \infty, \tag{B.1}$$

which is symmetric about 0, that is $\varphi(-x) = \varphi(x)$. The corresponding cumulative distribution function is denoted by

$$\Phi(x) = \int_{-\infty}^{x} \varphi(t)\,dt = \frac{1}{2}\left[\operatorname{erf}\left(\frac{x}{\sqrt{2}}\right) + 1\right] \tag{B.2}$$

for $-\infty < x < \infty$. Because of symmetry, we have

$$\Phi(x) + \Phi(-x) = 1, \qquad \Phi(0) = \tfrac{1}{2}.$$

The tail behaviour of $\Phi(x)$ is regulated by the following inequalities:

$$\frac{\varphi(x)}{x} - \frac{\varphi(x)}{x^3} < 1 - \Phi(x) < \frac{\varphi(x)}{x}, \qquad \text{if } x > 0. \tag{B.3}$$

The transformed variable $Y = \mu + \sigma X$ is said to be normally distributed with parameters μ and σ^2 for any $\mu \in \mathbb{R}$ and $\sigma \in \mathbb{R}^+$. In this case the notation $Y \sim N(\mu, \sigma^2)$ is used. The density function of Y at y is

$$\frac{1}{\sqrt{2\pi}\,\sigma} \exp\left[-\frac{1}{2}\left(\frac{y-\mu}{\sigma}\right)^2\right], \qquad -\infty < y < \infty. \tag{B.4}$$

The characteristic function, the moment generating function and the cumulant generating function of Y are

$$\Psi_Y(t) = \mathbb{E}\left\{e^{itY}\right\} = \exp\left(i\mu t - \tfrac{1}{2}\sigma^2 t^2\right), \tag{B.5}$$

$$M_Y(t) = \mathbb{E}\left\{e^{tY}\right\} = \exp\left(\mu t + \tfrac{1}{2}\sigma^2 t^2\right), \tag{B.6}$$

$$K_Y(t) = \log M_Y(t) = \mu t + \tfrac{1}{2}\sigma^2 t^2, \tag{B.7}$$

respectively, and

$$\mathbb{E}\{Y\} = \mu, \tag{B.8}$$

$$\mathbb{E}\{(Y - \mu)^k\} = \begin{cases} 0 & \text{if } k = 1, 3, \ldots, \\ (k-1)!! \, \sigma^k & \text{if } k = 2, 4, \ldots, \end{cases} \tag{B.9}$$

where the double factorial $n!!$ of an odd positive integer $n = 2m-1$ is

$$n!! = \prod_{j=1}^{m}(2j - 1) = \frac{(2m)!}{2^m \, m!}.$$

Proposition B.1 (Ellison, 1964) *If $Z \sim N(\mu, \sigma^2)$ and $W \sim \chi_q^2/q$ independently of Z, then for any c*

$$\mathbb{E}\left\{\Phi\left(Z + c\sqrt{W}\right)\right\} = \mathbb{P}\left\{T \leq \frac{c}{\sqrt{1 + \sigma^2}}\right\} \tag{B.10}$$

where T is a non-central t random variable with q degrees of freedom and non-centrality parameter $-\mu/\sqrt{1 + \sigma^2}$.

Corollary B.2 *If $Z \sim N(\mu, \sigma^2)$,*

$$\mathbb{E}\{\Phi(Z)\} = \Phi\left(\frac{\mu}{\sqrt{1 + \sigma^2}}\right). \tag{B.11}$$

Corollary B.3 *If $V \sim Gamma(s)$ and $T \sim t_{2s}$, then for any c*

$$\mathbb{E}\left\{\Phi\left(c\sqrt{V}\right)\right\} = \mathbb{P}\left\{T \leq c\sqrt{s}\right\}. \tag{B.12}$$

The following result is 'obvious' but, since no proof could be found in the literature, one is given here.

Proposition B.4 *For any choice of the constants a_1, b_1, a_2, b_2 such that $b_1 b_2 \neq 0$, there exist no constants a, b, c such that*

$$\Phi(a_1 + b_1 x)\, \Phi(a_2 + b_2 x) = c\, \Phi(a + b x), \quad \text{for all } x \in \mathbb{R}. \tag{B.13}$$

Proof If b_1 and b_2 have opposite signs, then the left side of (B.13) is the product of two positive functions, such that their product is 0 at $x \to \pm\infty$ and is positive otherwise, while the right side is monotone, leading to a contradiction. Hence b_1 and b_2 must have the same sign, and it is easy to see that this is the sign of b too. If b_1, b_2, b are all positive, consider $x \to -\infty$ (otherwise let $x \to \infty$) and, recalling (B.3), obtain that the ratio of the left and the right side of (B.13) as $x \to -\infty$ is

$$\frac{\varphi(a_1 + b_1 x)\varphi(a_2 + b_2 x)/(b_1 b_2 x^2)}{c\varphi(a + b x)/(-bx)} = \frac{\exp(\text{polynomial function of } x)}{-x}$$

which cannot converge to 1 as required by (B.13) to hold. QED

The bivariate normal distribution and related material

A bivariate continuous random variable $X = (X_1, X_2)$ is said to have a bivariate normal, or Gaussian, probability distribution with standardized marginals and correlation ρ if its density function at $(x_1, x_2) \in \mathbb{R}^2$ is

$$\varphi_B(x_1, x_2; \rho) = \frac{1}{2\pi(1-\rho^2)} \exp\left[-\frac{1}{2(1-\rho^2)}(x_1^2 - 2\rho x_1 x_2 + x_2^2)\right] \quad (\text{B.14})$$

for some $-1 < \rho < 1$. Each marginal component has density function of type (B.1), i.e., it is of $N(0, 1)$ type. The mean vector and the covariance matrix of X are

$$\mathbb{E}\{X\} = \begin{pmatrix} 0 \\ 0 \end{pmatrix}, \qquad \text{var}\{X\} = \begin{pmatrix} 1 & \rho \\ \rho & 1 \end{pmatrix}. \quad (\text{B.15})$$

Evaluation of the joint distribution function of X, that is

$$\Phi_B(x, y; \rho) = \mathbb{P}\{X_1 \leq h, X_2 \leq k\} = \int_{-\infty}^{h} \int_{-\infty}^{k} \varphi_B(x_1, x_2; \rho)\, dx_2\, dx_1, \quad (\text{B.16})$$

is not feasible in explicit form, except for some special cases such as the quadrant probability

$$\mathbb{P}\{X_1 \leq 0, X_2 \leq 0\} = \frac{1}{4} + \frac{\arcsin\rho}{2\pi} = \frac{\arccos(-\rho)}{2\pi}, \quad (\text{B.17})$$

amd in general we must resort to numerical methods. Owen (1956) has re-expressed (B.16) in terms of (B.2) and the auxiliary function

$$T(h, a) = \frac{1}{2\pi} \int_0^a \frac{\exp\{-\frac{1}{2}h^2(1+x^2)\}}{1+x^2}\, dx$$

$$= \frac{\arctan a}{2\pi} - \frac{1}{2\pi} \int_0^h \int_0^{ax} \exp\left[-\frac{1}{2}(x^2 + y^2)\right] dy\, dx$$

for $h, a \in \mathbb{R}$, arriving at

$$\mathbb{P}\{X_1 \leq h, X_2 \leq k\} = \frac{1}{2}\{\Phi(h) + \Phi(k)\} - T\left(h, \frac{k - \rho h}{h\sqrt{1-\rho^2}}\right)$$

$$-T\left(k, \frac{h - \rho k}{k\sqrt{1-\rho^2}}\right) - a(h, k) \quad (\text{B.18})$$

where

$$a(h, k) = \begin{cases} 0 & \text{if } hk > 0, \text{ or if } hk = 0 \text{ and } k \text{ or } k > 0, \text{ or if both} = 0, \\ \frac{1}{2} & \text{if } hk < 0, \text{ or if } hk = 0 \text{ and } h \text{ or } k < 0. \end{cases}$$

The function $T(h, a)$ enjoys several formal properties, namely

$$
\begin{aligned}
T(h, 0) &= 0, \\
T(0, a) &= (2\pi)^{-1} \arctan a, \\
T(h, -a) &= -T(h, a), \\
T(-h, a) &= T(h, a), \\
2T(h, 1) &= \Phi(h)\,\Phi(-h), \\
T(h, \infty) &= \begin{cases} \frac{1}{2}\{1 - \Phi(x)\} & \text{if } h \geq 0, \\ \frac{1}{2}\Phi(x) & \text{if } h \leq 0, \end{cases} \\
T(h, a) &= \tfrac{1}{2}\Phi(h) + \tfrac{1}{2}\Phi(ah) - \Phi(h)\Phi(ah) \\
&\quad -T(ah, 1/a) - \begin{cases} 0 & \text{if } a \geq 0, \\ \frac{1}{2} & \text{if } a < 0, \end{cases}
\end{aligned}
\tag{B.19}
$$

which are helpful in various ways, for instance reduction of its numerical evaluation to the case $0 < a < 1$.

Numerical tables of this function $T(h, a)$ for $0 < a < 1$ have been provided by Owen (1956) and more extensively by Owen (1957). Nowadays one would rather make use of a computer routine.

The monograph of Owen (1957) includes in addition a vast collection of formal results connected to the functions φ, Φ and T. Since this monograph is not commonly accessible, we reproduce here a few results of more direct relevance to our development, especially of Chapter 2. For arbitrary real numbers a and b,

$$
\int \varphi(x)\,\Phi(bx)\,\mathrm{d}x = -T(x, b) + \tfrac{1}{2}\Phi(x) + c,
\tag{B.20}
$$

$$
\begin{aligned}
\int \varphi(x)\,\Phi(a + bx)\,\mathrm{d}x &= T\left(x, \frac{a}{x\sqrt{1 + b^2}}\right) + T\left(\frac{a}{\sqrt{1 + b^2}}, \frac{x\sqrt{1 + b^2}}{a}\right) \\
&\quad - T\left(x, \frac{a + bx}{x}\right) - T\left(\frac{a}{\sqrt{1 + b^2}}, \frac{ab + x(1 + b^2)}{a}\right) \\
&\quad + \Phi(x)\,\Phi\left(\frac{a}{\sqrt{1 + b^2}}\right) + c,
\end{aligned}
\tag{B.21}
$$

$$
\begin{aligned}
\int x\,\varphi(x)\,\Phi(a + bx)\,\mathrm{d}x &= \frac{b}{\sqrt{1 + b^2}}\,\varphi\left(\frac{a}{\sqrt{1 + b^2}}\right)\Phi\left(x\sqrt{1 + b^2} + \frac{ab}{\sqrt{1 + b^2}}\right) \\
&\quad - \Phi(a + bx)\,\varphi(x) + c,
\end{aligned}
\tag{B.22}
$$

$$
\int_{-\infty}^{0} \varphi(x)\,\Phi(a + bx)\,\mathrm{d}x = \tfrac{1}{2}\,\Phi\left(\frac{a}{\sqrt{1 + b^2}}\right) - T\left(\frac{a}{\sqrt{1 + b^2}}, b\right),
\tag{B.23}
$$

$$
\int_{-\infty}^{\infty} \varphi(x)\,\Phi(a + bx)\,\mathrm{d}x = \Phi\left(\frac{a}{\sqrt{1 + b^2}}\right),
\tag{B.24}
$$

$$\int_h^k \varphi(x)\,\Phi(a+bx)\,\mathrm{d}x = \int_{-\infty}^{-\frac{a}{\sqrt{b^2+1}}} \varphi(x)\,\Phi(k\,\sqrt{b^2+1}+bx)\,\mathrm{d}x$$

$$- \int_{-\infty}^{-\frac{a}{\sqrt{b^2+1}}} \varphi(x)\,\Phi(h\,\sqrt{b^2+1}+bx)\,\mathrm{d}x, \quad \text{(B.25)}$$

$$\int_{-\infty}^{\infty} \{\Phi(ax)\}^2\,\{\varphi(x)\}^n\,\mathrm{d}x = \left(\pi - \arccos\frac{a^2}{n+a^2}\right) n^{-1/2}\,(2\pi)^{-(n+1)/2}$$

$$(n>0), \quad \text{(B.26)}$$

$$\int_{-\infty}^{\infty} \{\Phi(ax)\}^3\,\{\varphi(x)\}^n\,\mathrm{d}x = \tfrac{1}{2}\left(2\pi - 3\arccos\frac{a^2}{n+a^2}\right) n^{-1/2}\,(2\pi)^{-(n+1)/2}$$

$$(n>0), \quad \text{(B.27)}$$

$$\int_{-\infty}^{\infty} \{\Phi(ax+b)\}^2\,\varphi(x)\,\mathrm{d}x = \Phi\left(\frac{a}{\sqrt{1+b^2}}\right) - 2\,T\left(\frac{a}{\sqrt{1+b^2}},\frac{1}{\sqrt{1+2b^2}}\right).$$

$$\text{(B.28)}$$

The multivariate normal distribution

If Σ is a $d \times d$ symmetric positive-definite matrix and μ is a d-vector, we say that

$$\frac{1}{(2\pi)^{d/2}\,\det(\Sigma)^{1/2}}\,\exp\left(-\tfrac{1}{2}(x-\mu)^\top\Sigma^{-1}(x-\mu)\right), \qquad x \in \mathbb{R}^d \qquad \text{(B.29)}$$

is the d-dimensional normal density with parameters μ and Σ, although formally only the set of non-replicated values of Σ must be regarded as parameter components. The notation $\varphi_d(x;\Sigma)$ denotes this function when $\mu = 0$, so that (B.29) equals $\varphi_d(x-\mu;\Sigma)$. The corresponding distribution function is denoted $\Phi_d(x-\mu;\Sigma)$.

If X is a continuous random variable with density (B.29), we write $X \sim N_d(\mu,\Sigma)$. For this distribution of X, $a + A^\top X \sim N_p(a + A^\top\mu, A^\top\Sigma A)$, if a is a p-vector and A is a full-rank $d \times p$ matrix. The mean value, the variance matrix and the moment generating function of X are as follows:

$$\mathbb{E}\{X\} = \mu, \qquad \operatorname{var}\{X\} = \Sigma,$$
$$M(t) = \mathbb{E}\{\exp(t^\top X)\} = \exp\left(t^\top\mu + \tfrac{1}{2}t^\top\Sigma t\right).$$

There exist many distributional results for quadratic forms of X. Here we recall only the basic one, that is, $(X-\mu)^\top\Sigma^{-1}(X-\mu) \sim \chi_d^2$. Additional results are given in standard accounts such as the books of Rao (1973) and Mardia *et al.* (1979).

Appendix C

Notions on likelihood inference

Our notation and terminology related to likelihood inference are quite standard, but for completeness and to avoid ambiguities we recall briefly the essential concepts. Required regularity conditions are assumed to hold without more detailed specification. For a more detailed treatment the reader may wish to refer to a dedicated text; that of Azzalini (1996) presents the material at a level more than adequate for the requirements of this book and with a similar conception.

Consider a random variable Y whose probability distribution belongs to a parametric family whose elements are indexed by the parameter θ, where $\theta \in \mathbb{R}^k$; it is assumed that the parametric family is identifiable. Denote by $f(y; \theta)$ the density function of Y; if Y is a discrete random variable, the term 'density function' is used in a generalized sense to refer to the probability function. An important situation occurs when Y is n-dimensional with independent and identically distributed components; in this case the density function at $y = (y_1, \ldots, y_n)^\top$ is of the form

$$f(y; \theta) = \prod_{i=1}^{n} f_0(y_i; \theta)$$

where f_0 denotes the density function of a single component of Y.

If the observation y has been made on Y, then the likelihood function for θ is defined as

$$L(\theta) = c \, f(y; \theta) \qquad (C.1)$$

where c is an arbitrary positive constant which may depend on y, but not on θ. Often $c = 1$ is taken. In some cases, if we want to stress its dependence on y, we write $L(\theta; y)$ instead of $L(\theta)$; the same specification may be used for other functions to be introduced next. It is equivalent, and usually more convenient, to consider the log-likelihood function

$$\ell(\theta) = \log L(\theta) = \text{constant} + \log f(y; \theta). \qquad (C.2)$$

237

The criterion of maximum likelihood estimation operates by maximizing $L(\theta)$, or equivalently $\ell(\theta)$, with respect to θ. A value θ selected in this way, denoted $\hat{\theta}$, is called a *maximum likelihood estimate* (MLE). In many cases, this value is unique, or at least is believed to be unique; this motivates the common use of the phrase 'the MLE' instead of 'a MLE'. Usually, $\hat{\theta}$ is computed by solving the set of likelihood equations

$$s(\theta) = 0 \qquad\qquad (C.3)$$

where the k-valued function

$$s(\theta) = \frac{\mathrm{d}}{\mathrm{d}\theta}\ell(\theta) \qquad\qquad (C.4)$$

is called the *score function*. Since in most cases the likelihood equations are non-linear, their solutions can be accomplished only via numerical methods. Obviously, the sole fact that a point θ is a solution of (C.3) does not imply that it is the MLE; among the solutions of (C.3), $\hat{\theta}$ is the one which corresponds to the global maximum of $\ell(\theta)$.

Under regularity conditions, a Taylor series expansion of $\ell(\theta)$ around the point $\hat{\theta}$ gives the local approximation

$$\ell(\theta) = \ell(\hat{\theta}) - \tfrac{1}{2}(\theta - \hat{\theta})^{\top}\,\mathcal{J}(\hat{\theta})\,(\theta - \hat{\theta}) + \cdots \qquad\qquad (C.5)$$

where

$$\mathcal{J}(\hat{\theta}) = -\frac{\mathrm{d}}{\mathrm{d}\theta^{\top}}\,s(\theta)\bigg|_{\theta=\hat{\theta}} = -\frac{\mathrm{d}^2}{\mathrm{d}\theta\,\mathrm{d}\theta^{\top}}\,\ell(\theta)\bigg|_{\theta=\hat{\theta}} \qquad\qquad (C.6)$$

is called the *observed Fisher information*, which is a positive-definite $k \times k$ matrix. The remainder term of (C.5) is null when Y is of Gaussian type and θ is a linear function of $\mathbb{E}\{Y\}$. A connected quantity is the *expected Fisher information*

$$\mathcal{I}(\theta) = \mathbb{E}\left\{-\frac{\mathrm{d}}{\mathrm{d}\theta^{\top}}\,s(\theta; Y)\right\} = \mathbb{E}\{s(\theta; Y)\,s(\theta; Y)^{\top}\}, \qquad\qquad (C.7)$$

which in usual circumstances is a positive-definite matrix.

The study of formal properties of the MLE is possible in an exact form only for a limited set of cases. In general we must resort to some form of approximation, typically produced by an asymptotic argument. The basic situation is when the components of Y are n independent and identically distributed random variables. In this case it can be shown that, under fairly general regularity conditions, as $n \to \infty$

$$\hat{\theta} \xrightarrow{\ \mathrm{p}\ } \theta\,, \qquad\qquad (C.8)$$

$$\sqrt{n}(\hat{\theta} - \theta) \xrightarrow{\ \mathrm{d}\ } \mathrm{N}_k(0, \mathcal{I}_1(\theta)^{-1})\,, \qquad\qquad (C.9)$$

where $\mathcal{I}_1(\theta)$ denotes the expected Fisher information for a single component of Y; hence $n\,\mathcal{I}_1(\theta) = \mathcal{I}(\theta)$.

Outside the case of independent and identically distributed observations, a completely general statement is not feasible. In most cases, however, it can be proved that an approximation to the distribution the MLE is given by either of

$$\hat{\theta} - \theta \,\dot{\sim}\, N_k(0, \mathcal{I}(\theta)^{-1}), \qquad \hat{\theta} - \theta \,\dot{\sim}\, N_k(0, \mathcal{J}(\hat{\theta})^{-1}). \tag{C.10}$$

Taking the square root of the diagonal elements of $\mathcal{J}(\hat{\theta})^{-1}$ we obtain standard errors for $\hat{\theta}$; alternatively, standard errors can be obtained starting from $\mathcal{I}(\theta)^{-1}$ evaluated at $\theta = \hat{\theta}$. These two variant forms of standard errors tend to be numerically close, and in fact they can be shown to be exactly equal when $f(y; \theta)$ belongs to a regular exponential family.

A related distributional result is that, if θ_0 denotes the true parameter value and standard asymptotic theory holds, then

$$D(\theta_0) = 2\{\ell(\hat{\theta}) - \ell(\theta_0)\} \xrightarrow{\mathrm{d}} \chi_k^2 \tag{C.11}$$

holds asymptotically for the likelihood ratio test $D(\theta_0)$, hence allowing hypothesis testing for the parameter value. In addition, by exploiting the duality between hypothesis testing and interval estimation, we can use the result to construct confidence regions; this is more easily obtained by use of the *deviance function*

$$D(\theta) = 2\{\ell(\hat{\theta}) - \ell(\theta)\}, \tag{C.12}$$

briefly called 'deviance', such that $D(\theta) \geq 0$ and $D(\hat{\theta}) = 0$. From (C.5), we can approximate $D(\theta)$ in a neighborhood of $\hat{\theta}$ by a quadratic function of θ. For linear models under assumption of normality of the error terms, the function is exactly quadratic. The set

$$C(\theta) = \{\theta : 0 \leq D(\theta) \leq q_\alpha\}, \tag{C.13}$$

where q_α denotes the α-level upper quantile of the χ_k^2 distribution, represents a confidence region of approximate confidence level $1 - \alpha$.

Often the parameter can be split into two components, $\theta = (\psi, \lambda)$, where ψ denotes the component of interest, and λ is a nuisance parameter. It is then useful to introduce the profile log-likelihood

$$\ell^*(\psi) = \ell\big(\psi, \hat{\lambda}(\psi)\big) \tag{C.14}$$

where $\hat{\lambda}(\psi)$ denotes the value of λ which maximizes the likelihood when ψ is fixed at the chosen value. Obviously, if $\hat{\theta} = (\hat{\psi}, \hat{\lambda})$, then the maximum of ℓ^* occurs at $\psi = \hat{\psi}$.

Hypothesis testing and interval estimation for ψ can be accomplished on the basis of ℓ^*, using it similarly to ℓ. If the true value of ψ is ψ_0 and standard asymptotic theory holds, then

$$D(\psi_0) = 2\{\ell^*(\hat{\psi}) - \ell^*(\psi_0)\} \xrightarrow{\text{d}} \chi^2_h \qquad (\text{C.15})$$

where $h = \dim(\psi)$. Similarly to (C.12), the deviance function

$$D(\psi) = 2\{\ell^*(\hat{\psi}) - \ell^*(\psi)\} \qquad (\text{C.16})$$

can be used to construct a confidence region for ψ, using χ^2_h as the reference distribution.

References

Abe, T. and Pewsey, A. 2011. Sine-skewed circular distributions. *Statist. Papers*, **52**, 683–707. [210]

Adcock, C. J. 2004. Capital asset pricing in UK stocks under the multivariate skew-normal distribution. Chap. 11, pages 191–204 of: Genton, M. G. (ed.), *Skew-elliptical Distributions and their Applications: A Journey Beyond Normality*. Boca Raton, FL: Chapman & Hall/CRC. [159]

Adcock, C. J. 2007. Extensions of Stein's lemma for the skew-normal distribution. *Commun. Statist. Theory Methods*, **36**, 1661–1671. [163, 200]

Adcock, C. J. 2010. Asset pricing and portfolio selection based on the multivariate extended skew-Student-*t* distribution. *Ann. Oper. Res.*, **176**, 221–234. [183, 186]

Adcock, C. J. and Shutes, K. 1999. Portfolio selection based on the multivariate-skew normal distribution. Pages 167–177 of: Skulimowski, A. M. J. (ed.), *Financial Modelling*. Kraków: Progress and Business Publishers. Available in 2001. [142, 158, 186]

Aigner, D. J., Lovell, C. A. K., and Schmidt, P. 1977. Formulation and estimation of stochastic frontier production function model. *J. Economet.*, **6**, 21–37. [91]

Aitchison, J. 1986. *The Statistical Analysis of Compositional Data*. London: Chapman & Hall. [210, 211]

Anděl, J., Netuka, I., and Zvára, K. 1984. On threshold autoregressive processes. *Kybernetika*, **20**, 89–106. Prague: Academia. [43]

Arellano-Valle, R. B. 2010. The information matrix of the multivariate skew-*t* distribution. *Metron*, **LXVIII**, 371–386. [180]

Arellano-Valle, R. B. and Azzalini, A. 2006. On the unification of families of skew-normal distributions. *Scand. J. Statist.*, **33**, 561–574. [200, 201]

Arellano-Valle, R. B. and Azzalini, A. 2008. The centred parametrization for the multivariate skew-normal distribution. *J. Multiv. Anal.*, **99**, 1362–1382. Corrigendum: vol. 100 (2009), p. 816. [146, 149]

Arellano-Valle, R. B. and Azzalini, A. 2013. The centred parameterization and related quantities of the skew-*t* distribution. *J. Multiv. Anal.*, **113**, 73–90. Available online 12 June 2011. [114, 180]

Arellano-Valle, R. B. and del Pino, G. E. 2004. From symmetric to asymmetric distributions: a unified approach. Chap. 7, pages 113–130 of: Genton, M. G. (ed.), *Skew-elliptical Distributions and their Applications: A Journey Beyond Normality*. Boca Raton, FL: Chapman & Hall/CRC. [14]

241

Arellano-Valle, R. B. and Genton, M. G. 2005. On fundamental skew distributions. *J. Multiv. Anal.*, **96**, 93–116. [14, 23, 200]

Arellano-Valle, R. B. and Genton, M. G. 2007. On the exact distribution of linear combinations of order statistics from dependent random variables. *J. Multiv. Anal.*, **98**, 1876–1894. Corrigendum: **99** (2008) 1013. [203]

Arellano-Valle, R. B. and Genton, M. G. 2010a. An invariance property of quadratic forms in random vectors with a selection distribution, with application to sample variogram and covariogram estimators. *Ann. Inst. Statist. Math.*, **62**, 363–381. [14]

Arellano-Valle, R. B. and Genton, M. G. 2010b. Multivariate extended skew-*t* distributions and related families. *Metron*, **LXVIII**, 201–234. [183, 184, 194]

Arellano-Valle, R. B. and Genton, M. G. 2010c. Multivariate unified skew-elliptical distributions. *Chil. J. Statist.*, **1**, 17–33. [201]

Arellano-Valle, R. B. and Richter, W.-D. 2012. On skewed continuous $l_{n,p}$-symmetric distributions. *Chil. J. Statist.*, **3**, 195–214. [212, 213]

Arellano-Valle, R. B., del Pino, G., and San Martín, E. 2002. Definition and probabilistic properties of skew-distributions. *Statist. Probab. Lett.*, **58**, 111–121. [14]

Arellano-Valle, R. B., Gómez, H. W., and Quintana, F. A. 2004. A new class of skew-normal distributions. *Commun. Statist. Theory Methods*, **33**, 1465–1480. [48]

Arellano-Valle, R. B., Bolfarine, H., and Lachos, V. H. 2005a. Skew-normal linear mixed models. *J. Data Science*, **3**, 415–438. [94, 219]

Arellano-Valle, R. B., Gómez, H. W., and Quintana, F. A. 2005b. Statistical inference for a general class of asymmetric distributions. *J. Statist. Plann. Inference*, **128**, 427–443. [22]

Arellano-Valle, R. B., Branco, M. D., and Genton, M. G. 2006. A unified view on skewed distributions arising from selections. *Canad. J. Statist.*, **34**, 581–601. [14, 22]

Arellano-Valle, R. B., Bolfarine, H., and Lachos, V. H. 2007. Bayesian inference for skew-normal linear mixed models. *J. Appl. Statist.*, **34**, 663–682. [219]

Arellano-Valle, R. B., Genton, M. G., and Loschi, R. H. 2009. Shape mixtures of multivariate skew-normal distributions. *J. Multiv. Anal.*, **100**, 91–101. [49]

Arellano-Valle, R. B., Contreras-Reyes, J. E., and Genton, M. G. 2013. Shannon entropy and mutual information for multivariate skew-elliptical distributions. *Scand. J. Statist.*, **40**, 42–62. Available online 27 February 2012 (corrected 4 April 2012). [142]

Arnold, B. C. and Beaver, R. J. 2000a. Hidden truncation models. *Sankhyā, ser. A*, **62**, 22–35. [158]

Arnold, B. C. and Beaver, R. J. 2000b. The skew-Cauchy distribution. *Statist. Probab. Lett.*, **49**, 285–290. [190, 194]

Arnold, B. C. and Beaver, R. J. 2002. Skewed multivariate models related to hidden truncation and/or selective reporting (with discussion). *Test*, **11**, 7–54. [14]

Arnold, B. C. and Lin, G. D. 2004. Characterizations of the skew-normal and generalized chi distributions. *Sankhyā*, **66**, 593–606. [50]

Arnold, B. C., Beaver, R. J., Groeneveld, R. A., and Meeker, W. Q. 1993. The non-truncated marginal of a truncated bivariate normal distribution. *Psychometrika*, **58**, 471–478. [43, 87]

Arnold, B. C., Castillo, E., and Sarabia, J. M. 2002. Conditionally specified multivariate skewed distributions. *Sankhyā, ser. A*, **64**, 206–226. [23]

Azzalini, A. 1985. A class of distributions which includes the normal ones. *Scand. J. Statist.*, **12**, 171–178. [11, 43, 71, 72]

Azzalini, A. 1986. Further results on a class of distributions which includes the normal ones. *Statistica*, **XLVI**, 199–208. [11, 43, 101, 116, 123]

Azzalini, A. 1996. *Statistical Inference Based on the Likelihood*. London: Chapman & Hall. [237]

Azzalini, A. 2001. A note on regions of given probability of the skew-normal distribution. *Metron*, **LIX**, 27–34. [161]

Azzalini, A. 2005. The skew-normal distribution and related multivariate families (with discussion). *Scand. J. Statist.*, **32**, 159–188 (C/R 189–200). [44]

Azzalini, A. 2012. Selection models under generalized symmetry settings. *Ann. Inst. Statist. Math.*, **64**, 737–750. Available online 5 March 2011. [17, 23]

Azzalini, A. and Arellano-Valle, R. B. 2013. Maximum penalized likelihood estimation for skew-normal and skew-*t* distributions. *J. Statist. Plann. Inference*, **143**, 419–433. Available online 30 June 2012. [80, 82, 112]

Azzalini, A. and Bacchieri, A. 2010. A prospective combination of phase II and phase III in drug development. *Metron*, **LXVIII**, 347–369. [200, 225]

Azzalini, A. and Capitanio, A. 1999. Statistical applications of the multivariate skew normal distribution. *J. R. Statist. Soc., ser. B*, **61**, 579–602. Full version of the paper at arXiv.org:0911.2093. [11, 17, 71, 141, 145, 165, 175]

Azzalini, A. and Capitanio, A. 2003. Distributions generated by perturbation of symmetry with emphasis on a multivariate skew *t* distribution. *J. R. Statist. Soc., ser. B*, **65**, 367–389. Full version of the paper at arXiv.org:0911.2342. [11, 105, 111, 175, 178, 179, 193, 194]

Azzalini, A. and Chiogna, M. 2004. Some results on the stress–strength model for skew-normal variates. *Metron*, **LXII**, 315–326. [225]

Azzalini, A. and Dalla Valle, A. 1996. The multivariate skew-normal distribution. *Biometrika*, **83**, 715–726. [140, 165]

Azzalini, A. and Genton, M. G. 2008. Robust likelihood methods based on the skew-*t* and related distributions. *Int. Statist. Rev.*, **76**, 106–129. [112, 116, 145]

Azzalini, A. and Regoli, G. 2012a. Some properties of skew-symmetric distributions. *Ann. Inst. Statist. Math.*, **64**, 857–879. Available online 9 September 2011. [11, 19, 175, 189]

Azzalini, A. and Regoli, G. 2012b. The work of Fernando de Helguero on non-normality arising from selection. *Chil. J. Statist.*, **3**, 113–129. [46]

Azzalini, A., Dal Cappello, T., and Kotz, S. 2003. Log-skew-normal and log-skew-*t* distributions as model for family income data. *J. Income Distrib.*, **11**, 12–20. [54]

Azzalini, A., Genton, M. G., and Scarpa, B. 2010. Invariance-based estimating equations for skew-symmetric distributions. *Metron*, **LXVIII**, 275–298. [55, 206]

Balakrishnan, N. 2002. Comment to a paper by B. C. Arnold & R. Beaver. *Test*, **11**, 37–39. [201, 202]

Balakrishnan, N. and Scarpa, B. 2012. Multivariate measures of skewness for the skew-normal distribution. *J. Multiv. Anal.*, **104**, 73–87. [141]

Basso, R. M., Lachos, V. H., Cabral, C. R. B., and Ghosh, P. 2010. Robust mixture modeling based on scale mixtures of skew-normal distributions. *Comp. Statist. Data An.*, **54**, 2926–2941. [221]

Bayes, C. L. and Branco, M. D. 2007. Bayesian inference for the skewness parameter of the scalar skew-normal distribution. *Brazilian J. Probab. Stat.*, **21**, 141–163. [83, 84]

Bazán, J. L., Branco, M. D., and Bolfarine, H. 2006. A skew item response model. *Bayesian Anal.*, **1**, 861–892. [227]

Behboodian, J., Jamalizadeh, A., and Balakrishnan, N. 2006. A new class of skew-Cauchy distributions. *Statist. Probab. Lett.*, **76**, 1488–1493. [120]

Berlik, S. 2006. *Directed Evolutionary Algorithms*. Dissertation zur Erlangung des Grades eines Doktors der Naturwissenschaften, Universität Dortmund, Fachbereich Informatik, Dortmund. [218]

Birnbaum, Z. W. 1950. Effect of linear truncation on a multinormal population. *Ann. Math. Statist.*, **21**, 272–279. [42]

Bolfarine, H., Montenegro, L. C., and Lachos, V. H. 2007. Influence diagnostics for skew-normal linear mixed models. *Sankhyā*, **69**, 648–670. [220]

Box, G. P. and Tiao, G. C. 1973. *Bayesian Inference in Statistical Analysis*. New York: Addison-Wesley. [95]

Branco, M. D. and Dey, D. K. 2001. A general class of multivariate skew-elliptical distributions. *J. Multiv. Anal.*, **79**, 99–113. [104, 175, 178]

Branco, M. D. and Dey, D. K. 2002. Regression model under skew elliptical error distribution. *J. Math. Sci. (New Series), Delhi*, **1**, 151–168. [111]

Cabral, C. R. B., Lachos, V. H., and Prates, M. O. 2012. Multivariate mixture modeling using skew-normal independent distributions. *Comp. Statist. Data An.*, **56**, 126–142. [221]

Cabras, S. and Castellanos, M. E. 2009. Default Bayesian goodness-of-fit tests for the skew-normal model. *J. Appl. Statist.*, **36**, 223–232. [87]

Cabras, S., Racugno, W., Castellanos, M. E., and Ventura, L. 2012. A matching prior for the shape parameter of the skew-normal distribution. *Scand. J. Statist.*, **39**, 236–247. [84]

Callegaro, A. and Iacobelli, S. 2012. The Cox shared frailty model with log-skew-normal frailties. *Statist. Model.*, **12**, 399–418. [228]

Canale, A. 2011. Statistical aspects of the scalar extended skew-normal distribution. *Metron*, **LXIX**, 279–295. [55, 87]

Capitanio, A. 2010. On the approximation of the tail probability of the scalar skew-normal distribution. *Metron*, **LXVIII**, 299–308. [53]

Capitanio, A. 2012. *On the canonical form of scale mixtures of skew-normal distributions*. Available at arXiv.org:1207.0797. [123, 141, 175, 195]

Capitanio, A. and Pacillo, S. 2008. A Wald's test for conditional independence skew normal graphs. Pages 421–428 of: *Proceedings in Computational Statistics: CompStat 2008*. Heidelberg: Physica-Verlag. [158]

Capitanio, A., Azzalini, A., and Stanghellini, E. 2003. Graphical models for skew-normal variates. *Scand. J. Statist.*, **30**, 129–144. [87, 158]

Cappuccio, N., Lubian, D., and Raggi, D. 2004. MCMC Bayesian estimation of a skew-GED stochastic volatility model. *Studies in Nonlinear Dynamics and Econometrics*, **8**, Article 6. [101]

Carmichael, B. and Coën, A. 2013. Asset pricing with skewed-normal return. *Finance Res. Letters*, **10**, 50–57. Available online 1 February 2013. [159]

Carota, C. 2010. Tests for normality in classes of skew-*t* alternatives. *Statist. Probab. Lett.*, **80**, 1–8. [122]

Chai, H. S. and Bailey, K. R. 2008. Use of log-skew-normal distribution in analysis of continuous data with a discrete component at zero. *Statist. Med.*, **27**, 3643–3655. [54]

Chang, C.-H., Lin, J.-J., Pal, N., and Chiang, M.-C. 2008. A note on improved approximation of the binomial distribution by the skew-normal distribution. *Amer. Statist.*, **62**, 167–170. [215]

Chang, S.-M. and Genton, M. G. 2007. Extreme value distributions for the skew-symmetric family of distributions. *Commun. Statist. Theory Methods*, **36**, 1705–1717. [53, 122]

Chen, J. T. and Gupta, A. K. 2005. Matrix variate skew normal distributions. *Statistics*, **39**, 247–253. [212]

Chen, M.-H. 2004. Skewed link models for categorical response data. Chap. 8, pages 131–152 of: Genton, M. G. (ed.), *Skew-elliptical Distributions and their Applications: A Journey Beyond Normality*. Boca Raton, FL: Chapman & Hall/CRC. [226]

Chen, M.-H., Dey, D. K., and Shao, Q.-M. 1999. A new skewed link model for dichotomous quantal response data. *J. Amer. Statist. Assoc.*, **94**, 1172–1186. [226]

Chiogna, M. 1998. Some results on the scalar skew-normal distribution. *J. Ital. Statist. Soc.*, **7**, 1–13. [43, 51, 54]

Chiogna, M. 2005. A note on the asymptotic distribution of the maximum likelihood estimator for the scalar skew-normal distribution. *Stat. Meth. & Appl.*, **14**, 331–341. [72]

Chu, K. K., Wang, N., Stanley, S., and Cohen, N. D. 2001. Statistical evaluation of the regulatory guidelines for use of furosemide in race horses. *Biometrics*, **57**, 294–301. [160]

Churchill, E. 1946. Information given by odd moments. *Ann. Math. Statist.*, **17**, 244–246. [123]

Coelli, T. J., Prasada Rao, D. S., O'Donnell, C., and Battese, G. E. 2005. *An Introduction to Efficiency and Productivity Analysis*, 2nd edn. Berlin: Springer-Verlag. [91]

Contreras-Reyes, J. E. and Arellano-Valle, R. B. 2012. Kullback–Leibler divergence measure for multivariate skew-normal distributions. *Entropy*, **14**, 1606–1626. [142]

Copas, J. B. and Li, H. G. 1997. Inference for non-random samples (with discussion). *J. R. Statist. Soc., ser. B*, **59**, 55–95. [89]

Corns, T. R. A. and Satchell, S. E. 2007. Skew Brownian motion and pricing European options. *European J. Finance*, **13**, 523–544. [223]

Corns, T. R. A. and Satchell, S. E. 2010. Modelling conditional heteroskedasticity and skewness using the skew-normal distribution one-sided coverage intervals with survey data. *Metron*, **LXVIII**, 251–263. [224]

Cox, D. R. 1977. Discussion of 'Do robust estimators work with *real* data?' by Stephen M. Stigler. *Ann. Statist.*, **5**, 1083. [97]

Cox, D. R. 2006. *Principles of Statistical Inference*. Cambridge: Cambridge University Press. [69]

Cox, D. R. and Wermuth, N. 1996. *Multivariate Dependencies: Models, Analysis and Interpretation*. London: Chapman & Hall. [154]

Cramér, H. 1946. *Mathematical Methods of Statistics*. Princeton, NJ: Princeton University Press. [33, 61]

Dalla Valle, A. 1998. *La Distribuzione Normale Asimmetrica: Problematiche e Utilizzi nelle Applicazioni*. Tesi di dottorato, Dipartimento di Scienze Statistiche, Università di Padova, Padova, Italia. [56]

Dalla Valle, A. 2007. A test for the hypothesis of skew-normality in a population. *J. Statist. Comput. Simul.*, **77**, 63–77. [86]

de Helguero, F. 1909a. Sulla rappresentazione analitica delle curve abnormali. Pages 288–299 of: Castelnuovo, G. (ed.), *Atti del IV Congresso Internazionale dei Matematici (Roma, 6–11 Aprile 1908)*, vol. III, sezione III-B. Roma: R. Accademia dei Lincei. Available at http://www.mathunion.org/ICM/ICM1908.3/Main/icm1908.3.0288.0299.ocr.pdf. [44]

de Helguero, F. 1909b. Sulla rappresentazione analitica delle curve statistiche. *Giornale degli Economisti*, **XXXVIII, serie 2**, 241–265. [44]

De Luca, G. and Loperfido, N. M. R. 2004. A skew-in-mean GARCH model. Chap. 12, pages 205–222 of: Genton, M. G. (ed.), *Skew-elliptical Distributions and their Applications: A Journey Beyond Normality*. Boca Raton, FL: Chapman & Hall/CRC. [224]

De Luca, G., Genton, M. G., and Loperfido, N. 2005. A multivariate skew-GARCH model. *Adv. Economet.*, **20**, 33–57. [224]

Dharmadhikari, S. W. and Joag-dev, K. 1988. *Unimodality, Convexity, and Applications*. New York: Academic Press. [19, 189]

DiCiccio, T. J. and Monti, A. C. 2004. Inferential aspects of the skew exponential power distribution. *J. Amer. Statist. Assoc.*, **99**, 439–450. [101]

DiCiccio, T. J. and Monti, A. C. 2011. Inferential aspects of the skew *t*-distribution. *Quaderni di Statistica*, **13**, 1–21. [112]

Domínguez-Molina, J. A. and Rocha-Arteaga, A. 2007. On the infinite divisibility of some skewed symmetric distributions. *Statist. Probab. Lett.*, **77**, 644–648. [54]

Domínguez-Molina, J. A., González-Farías, G., and Ramos-Quiroga, R. 2004. Skew-normality in stochastic frontier analysis. Chap. 13, pages 223–242 of: Genton, M. G. (ed.), *Skew-elliptical Distributions and their Applications: A Journey Beyond Normality*. Boca Raton, FL: Chapman & Hall/CRC. [225]

Efron, B. 1981. Nonparametric standard errors and confidence intervals (with discussion). *Canad. J. Statist.*, **9**, 139–172. [55]

Elal-Olivero, D., Gómez, H. W., and Quintana, F. A. 2009. Bayesian modeling using a class of bimodal skew-elliptical distributions. *J. Statist. Plann. Inference*, **139**, 1484–1492. [213]

Elandt, R. C. 1961. The folded normal distribution: two methods of estimating parameters from moment. *Technometrics*, **3**, 551–562. [52]

Ellison, B. E. 1964. Two theorems for inferences about the normal distribution with applications in acceptance sampling. *J. Amer. Statist. Assoc.*, **59**, 89–95. [26, 233]

Fang, B. Q. 2003. The skew elliptical distributions and their quadratic forms. *J. Multiv. Anal.*, **87**, 298–314. [175, 193]

Fang, B. Q. 2005a. Noncentral quadratic forms of the skew elliptical variables. *J. Multiv. Anal.*, **95**, 410–430. [175]

Fang, B. Q. 2005b. The *t* statistic of the skew elliptical distributions. *J. Statist. Plann. Inference*, **134**, 140–157. [175]

Fang, B. Q. 2006. Sample mean, covariance and T^2 statistic of the skew elliptical model. *J. Multiv. Anal.*, **97**, 1675–1690. [175]

Fang, B. Q. 2008. Noncentral matrix quadratic forms of the skew elliptical variables. *J. Multiv. Anal.*, **99**, 1105–1127. [175]

Fang, K.-T. and Zhang, Y.-T. 1990. *Generalized Multivariate Analysis*. Berlin: Springer Verlag. [168]

Fang, K.-T., Kotz, S., and Ng, K. W. 1990. *Symmetric Multivariate and Related Distributions*. London: Chapman & Hall. [168]

Fechner, G. T. 1897. *Kollectivmasslehre*. Leipzig: Verlag von Wilhelm Engelmann. Published posthumously, completed and edited by G. F. Lipps. [21]

Fernández, C. and Steel, M. F. J. 1998. On Bayesian modeling of fat tails and skewness. *J. Amer. Statist. Assoc.*, **93**, 359–371. [22]

Firth, D. 1993. Bias reduction of maximum likelihood estimates. *Biometrika*, **80**, 27–38. Amendment: vol. 82, 667. [79]

Flecher, C., Allard, D., and Naveau, P. 2010. Truncated skew-normal distributions: moments, estimation by weighted moments and application to climatic data. *Metron*, **LXVIII**, 331–345. [52]

Forina, M., Armanino, C., Castino, M., and Ubigli, M. 1986. Multivariate data analysis as a discriminating method of the origin of wines. *Vitis*, **25**, 189–201. [59]

Frederic, P. 2011. Modeling skew-symmetric distributions using B-spline and penalties. *J. Statist. Plann. Inference*, **141**, 2878–2890. [204]

Frühwirth-Schnatter, S. and Pyne, S. 2010. Bayesian inference for finite mixtures of univariate and multivariate skew-normal and skew-*t* distributions. *Biostatistics*, **11**, 317–336. [221, 222]

Fung, T. and Seneta, E. 2010. Tail dependence for two skew *t* distributions. *Statist. Probab. Lett.*, **80**, 784–791. [193]

Genton, M. G. (ed.). 2004. *Skew-elliptical Distributions and their Applications: A Journey Beyond Normality*. Boca Raton, FL: Chapman & Hall/CRC. [186]

Genton, M. G. 2005. Discussion of 'The skew-normal'. *Scand. J. Statist.*, **32**, 189–198. [204]

Genton, M. G. and Loperfido, N. 2005. Generalized skew-elliptical distributions and their quadratic forms. *Ann. Inst. Statist. Math.*, **57**, 389–401. [11, 175]

Genton, M. G., He, L., and Liu, X. 2001. Moments of skew-normal random vectors and their quadratic forms. *Statist. Probab. Lett.*, **51**, 319–325. [142]

Ghizzoni, T., Roth, G., and Rudari, R. 2010. Multivariate skew-*t* approach to the design of accumulation risk scenarios for the flooding hazard. *Advances in Water Resources*, **33**, 1243–1255. [186]

Ghizzoni, T., Roth, G., and Rudari, R. 2012. Multisite flooding hazard assessment in the Upper Mississippi River. *J. Hydrology*, **412–413**, 101–113. [186]

Ghosh, P., Branco, M. D., and Chakraborty, H. 2007. Bivariate random effect model using skew-normal distribution with application to HIV–RNA. *Statist. Med.*, **26**, 1255–1267. [220]

Giorgi, E. 2012. *Indici non Parametrici per Famiglie Parametriche con Particolare Riferimento alla t Asimmetrica*. Tesi di laurea magistrale, Università di Padova. http://tesi.cab.unipd.it/40101/. [180]

González-Farías, G., Domínguez-Molina, J. A., and Gupta, A. K. 2004a. Additive properties of skew normal random vectors. *J. Statist. Plann. Inference*, **126**, 521–534. [200]

González-Farías, G., Domínguez-Molina, J. A., and Gupta, A. K. 2004b. The closed skew-normal distribution. Chap. 2, pages 25–42 of: Genton, M. G. (ed.), *Skew-elliptical Distributions and their Applications: A Journey Beyond Normality*. Boca Raton, FL: Chapman & Hall/CRC. [200]

Greco, L. 2011. Minimum Hellinger distance based inference for scalar skew-normal and skew-t distributions. *Test*, **20**, 120–137. [82]

Grilli, L. and Rampichini, C. 2010. Selection bias in linear mixed models. *Metron*, **LXVIII**, 309–329. [200]

Guolo, A. 2008. A flexible approach to measurement error correction in case-control studies. *Biometrics*, **64**, 1207–1214. [216]

Gupta, A. K. 2003. Multivariate skew t-distribution. *Statistics*, **37**, 359–363. [105, 178]

Gupta, A. K. and Huang, W.-J. 2002. Quadratic forms in skew normal variates. *J. Math. Anal. Appl.*, **273**, 558–564. [142]

Gupta, A. K. and Kollo, T. 2003. Density expansions based on the multivariate skew normal distribution. *Sankhyā*, **65**, 821–835. [216]

Gupta, A. K., Chang, F. C., and Huang, W.-J. 2002. Some skew-symmetric models. *Random Op. Stochast. Eq.*, **10**, 133–140. [120]

Gupta, A. K., González-Farías, G., and Domínguez-Molina, J. A. 2004. A multivariate skew normal distribution. *J. Multiv. Anal.*, **89**, 181–190. [200]

Gupta, R. C. and Brown, N. 2001. Reliability studies of the skew-normal distribution and its application to a strength–stress model. *Commun. Statist. Theory Methods*, **30**, 2427–2445. [225]

Gupta, R. C. and Gupta, R. D. 2004. Generalized skew normal model. *Test*, **13**, 501–524. [202]

Hallin, M. and Ley, C. 2012. Skew-symmetric distributions and Fisher information – a tale of two densities. *Bernoulli*, **18**, 747–763. [188]

Hampel, F. R., Rousseeuw, P. J., Ronchetti, E. M., and Stahel, W. A. 1986. *Robust Statistics: The Approach Based on Influence Functions*. New York: J. Wiley & Sons. [116]

Hansen, B. 1994. Autoregressive conditional density estimation. *Int. Econ. Rev.*, **35**, 705–730. [22]

Harrar, S. W. and Gupta, A. K. 2008. On matrix variate skew-normal distributions. *Statistics*, **42**, 179–184. [212]

Healy, M. J. R. 1968. Multivariate normal plotting. *Appl. Statist.*, **17**, 157–161. [144]

Heckman, J. J. 1976. The common structure of statistical models of truncation, sample selection and limited dependent variables, and a simple estimator for such models. *Ann. Econ. Soc. Meas.*, **5**, 475–492. [89, 90]

Henze, N. 1986. A probabilistic representation of the 'skew-normal' distribution. *Scand. J. Statist.*, **13**, 271–275. [43, 54]

Hernández-Sánchez, E. and Scarpa, B. 2012. A wrapped flexible generalized skew-normal model for a bimodal circular distribution of wind directions. *Chil. J. Statist.*, **3**, 131–143. [208]

Hill, M. A. and Dixon, W. J. 1982. Robustness in real life: a study of clinical laboratory data. *Biometrics*, **38**, 377–396. [96]

Hinkley, D. V. and Revankar, N. S. 1977. Estimation of the Pareto law from underreported data. *J. Economet.*, **5**, 1–11. [22]

Ho, H.-J. and Lin, T.-I. 2010. Robust linear mixed models using the skew *t* distribution with application to schizophrenia data. *Biometr. J.*, **52**, 449–469. [220]

Huang, W.-J. and Chen, Y.-H. 2007. Generalized skew-Cauchy distribution. *Statist. Probab. Lett.*, **77**, 1137–1147. [19]

Huber, P. J. 1981. *Robust Statistics*. New York: J. Wiley & Sons. [116]

Huber, P. J. and Ronchetti, E. M. 2009. *Robust Statistics*, 2nd edn. New York: J. Wiley & Sons. [118]

Jamalizadeh, A. and Balakrishnan, N. 2008. On order statistics from bivariate skew-normal and skew-t_ν distributions. *J. Statist. Plann. Inference*, **138**, 4187–4197. [202]

Jamalizadeh, A. and Balakrishnan, N. 2009. Order statistics from trivariate normal and t_ν-distributions in terms of generalized skew-normal and skew-t_ν distributions. *J. Statist. Plann. Inference*, **139**, 3799–3819. [202, 203]

Jamalizadeh, A. and Balakrishnan, N. 2010. Distributions of order statistics and linear combinations of order statistics from an elliptical distribution as mixtures of unified skew-elliptical distributions. *J. Multiv. Anal.*, **101**, 1412–1427. [201, 203]

Jamalizadeh, A., Khosravi, M., and Balakrishnan, N. 2009a. Recurrence relations for distributions of a skew-*t* and a linear combination of order statistics from a bivariate-*t*. *Comp. Statist. Data An.*, **53**, 847–852. [121]

Jamalizadeh, A., Mehrali, Y., and Balakrishnan, N. 2009b. Recurrence relations for bivariate *t* and extended skew-*t* distributions and an application to order statistics from bivariate *t*. *Comp. Statist. Data An.*, **53**, 4018–4027. [183, 186]

Jamshidi, A. A. and Kirby, M. J. 2010. Skew-radial basis function expansions for empirical modeling. *SIAM J. Sci. Comput.*, **31**, 4715–4743. [217]

Jara, A., Quintana, F., and San Martín, E. 2008. Linear mixed models with skew-elliptical distributions: a Bayesian approach. *Comp. Statist. Data An.*, **52**, 5033–5045. [220]

Javier, W. and Gupta, A. K. 2009. Mutual information for certain multivariate distributions. *Far East J. Theor. Stat.*, **29**, 39–51. [142]

Jiménez-Gamero, M. D., Alba-Fernández, V., Muñoz-García, J., and Chalco-Cano, Y. 2009. Goodness-of-fit tests based on empirical characteristic functions. *Comp. Statist. Data An.*, **53**, 3957–3971. [146]

Jones, M. C. 2001. A skew *t* distribution. Pages 269–278 of: Charalambides, C. A., Koutras, M. V., and Balakrishnan, N. (eds), *Probability and Statistical Models with Applications: A Volume in Honor of Theophilos Cacoullos*. London: Chapman & Hall. [106]

Jones, M. C. 2012. Relationship between distributions with certain symmetries. *Statist. Probab. Lett.*, **82**, 1737–1744. [21]

Jones, M. C. 2013. Generating distributions by transformation of scale. *Statist. Sinica*, to appear. [20, 21]

Jones, M. C. and Faddy, M. J. 2003. A skew extension of the *t*-distribution, with applications. *J. R. Statist. Soc., ser. B*, **65**, 159–174. [106, 108]

Jones, M. C. and Larsen, P. V. 2004. Multivariate distributions with support above the diagonal. *Biometrika*, **91**, 975–986. [107]

Kano, Y. 1994. Consistency property of elliptical probability density functions. *J. Multiv. Anal.*, **51**, 139–147. [107, 171]

Kim, H. J. 2002. Binary regression with a class of skewed t link models. *Commun. Statist. Theory Methods*, **31**, 1863–1886. [227]

Kim, H.-J. 2008. A class of weighted multivariate normal distributions and its properties. *J. Multiv. Anal.*, **99**, 1758–1771. [166]

Kim, H.-M. and Genton, M. G. 2011. Characteristic functions of scale mixtures of multivariate skew-normal distributions. *J. Multiv. Anal.*, **102**, 1105–1117. [51, 175]

Kim, H.-M. and Mallick, B. K. 2003. Moments of random vectors with skew t distribution and their quadratic forms. *Statist. Probab. Lett.*, **63**, 417–423. Corrigendum: vol. 79 (2009), 2098–2099. [178]

Kim, H.-M. and Mallick, B. K. 2004. A Bayesian prediction using the skew Gaussian distribution. *J. Statist. Plann. Inference*, **120**, 85–101. [224]

Kim, H.-M., Ha, E. and Mallick, B. K. 2004. Spatial prediction of rainfall using skew-normal processes. Chap. 16, pages 279–289 of: Genton, M. G. (ed.), *Skew-elliptical Distributions and their Applications: A Journey Beyond Normality*. Boca Raton, FL: Chapman & Hall/CRC. [224]

Kozubowski, T. J. and Nolan, J. P. 2008. Infinite divisibility of skew Gaussian and Laplace laws. *Statist. Probab. Lett.*, **78**, 654–660. [54]

Lachos, V. H., Ghosh, P., and Arellano-Valle, R. B. 2010a. Likelihood based inference for skew-normal independent linear mixed models. *Statist. Sinica*, **20**, 303–322. [175]

Lachos, V. H., Labra, F. V., Bolfarine, H., and Ghosh, P. 2010b. Multivariate measurement error models based on scale mixtures of the skew-normal distribution. *Statistics*, **44**, 541–556. Available online 28 October 2009. [179]

Lagos Álvarez, B. and Jiménez Gamero, M. D. 2012. A note on bias reduction of maximum likelihood estimates for the scalar skew t distribution. *J. Statist. Plann. Inference*, **142**, 608–612. Available online 8 September 2011. [112]

Lange, K. L., Little, R. J. A., and Taylor, J. M. G. 1989. Robust statistical modeling using the t-distribution. *J. Amer. Statist. Assoc.*, **84**, 881–896. [95]

Lauritzen, S. L. 1996. *Graphical Models*. Oxford: Oxford University Press. [154]

Leadbetter, M. R., Lindgren, G., and Rootzén, H. 1983. *Extremes and Related Properties of Random Sequences and Processes*. Berlin: Springer-Verlag. [55, 122]

Lee, S. and McLachlan, G. J. 2012. Finite mixtures of multivariate skew t-distributions: some recent and new results. *Statist. Comput.*, to appear. Available online 20 October 2012. [192]

Lee, S., Genton, M. G., and Arellano-Valle, R. B. 2010. Perturbation of numerical confidential data via skew-t distributions. *Manag. Sci.*, **56**, 318–333. [185]

Ley, C. and Paindaveine, D. 2010a. On Fisher information matrices and profile log-likelihood functions in generalized skew-elliptical models. *Metron*, **LXVIII**, 235–250. [180]

Ley, C. and Paindaveine, D. 2010b. On the singularity of multivariate skew-symmetric models. *J. Multiv. Anal.*, **101**, 1434–1444. [188]

Lin, G. D. and Stoyanov, J. 2009. The logarithmic skew-normal distributions are moment-indeterminate. *J. Appl. Prob.*, **46**, 909–916. [54]

Lin, T. I., 2009. Maximum likelihood estimation for multivariate skew normal mixture models. *J. Multiv. Anal.*, **100**, 257–265. [221]

Lin, T.-I. 2010. Robust mixture modeling using multivariate skew t distributions. *Statist. Comput.*, **20**, 343–356. [192, 221]

Lin, T.-I. and Lin, T.-C. 2011. Robust statistical modelling using the multivariate skew *t* distribution with complete and incomplete data. *Statist. Model.*, **11**, 253–277. [192]

Lin, T. I., Lee, J. C., and Hsieh, W. J. 2007a. Robust mixture modeling using the skew *t* distribution. *Statist. and Comput.*, **17**, 81–92. [221]

Lin, T. I., Lee, J. C., and Yen, S. Y. 2007b. Finite mixture modelling using the skew normal distribution. *Statist. Sinica*, **17**, 909–927. [94, 221]

Liseo, B. 1990. La classe delle densità normali sghembe: aspetti inferenziali da un punto di vista bayesiano. *Statistica*, **L**, 59–70. [77]

Liseo, B. and Loperfido, N. 2003. A Bayesian interpretation of the multivariate skew-normal distribution. *Statist. Probab. Lett.*, **61**, 395–401. [200]

Liseo, B. and Loperfido, N. 2006. A note on reference priors for the scalar skew-normal distribution. *J. Statist. Plann. Inference*, **136**, 373–389. [82, 83]

Loperfido, N. 2001. Quadratic forms of skew-normal random vectors. *Statist. Probab. Lett.*, **54**, 381–387. [141]

Loperfido, N. 2002. Statistical implications of selectively reported inferential results. *Statist. Probab. Lett.*, **56**, 13–22. [43]

Loperfido, N. 2008. Modelling maxima of longitudinal contralateral observations. *Test*, **17**, 370–380. [141]

Loperfido, N. 2010. Canonical transformations of skew-normal variates. *Test*, **19**, 146–165. [141]

Lysenko, N., Roy, P., and Waeber, R. 2009. Multivariate extremes of generalized skew-normal distributions. *Statist. Probab. Lett.*, **79**, 525–533. [23, 193]

Ma, Y. and Genton, M. G. 2004. Flexible class of skew-symmetric distributions. *Scand. J. Statist.*, **31**, 459–468. [50, 203, 204]

Ma, Y., Genton, M. G., and Tsiatis, A. A. 2005. Locally efficient semiparametric estimators for generalized skew-elliptical distributions. *J. Amer. Statist. Assoc.*, **100**, 980–989. [205]

Maddala, G. S. 2006. Limited dependent variables models. In: *Encyclopedia of Statistical Sciences*. New York: J. Wiley & Sons. [89, 90]

Malkovich, J. F. and Afifi, A. A. 1973. Measures of multivariate skewness and kurtosis with applications. *J. Amer. Statist. Assoc.*, **68**, 176–179. [138]

Marchenko, Y. V. and Genton, M. G. 2012. A Heckman selection-*t* model. *J. Amer. Statist. Assoc.*, **107**, 304–317. [185, 186]

Mardia, K. 1970. Measures of multivariate skewness and kurtosis with applications. *Biometrika*, **57**, 519–530. [132]

Mardia, K. V. 1974. Applications of some measures of multivariate skewness and kurtosis in testing normality and robustness studies. *Sankhyā, ser. B*, **36**, 115–128. [132, 174]

Mardia, K. V. and Jupp, P. E. 1999. *Directional Statistics*. New York: J. Wiley & Sons. [208]

Mardia, K. V., Kent, J. T., and Bibby, J. M. 1979. *Multivariate Analysis*. New York: Academic Press. [137]

Martínez, E. H., Varela, H., Gómez, H. W., and Bolfarine, H. 2008. A note on the likelihood and moments of the skew-normal distribution. *SORT*, **32**, 57–66. [54, 94]

Mateu-Figueras, G. and Pawlowsky-Glahn, V. 2007. The skew-normal distribution on the simplex. *Commun. Statist. Theory Methods*, **36**, 1787–1802. [211]

Mateu-Figueras, G., Pawlowsky-Glahn, V., and Barceló-Vidal, C. 2005. Additive logistic skew-normal on the simplex. *Stochast. Environ. Res. Risk Assess.*, **19**, 205–214. [211]

Mateu-Figueras, G., Puig, P., and Pewsey, A. 2007. Goodness-of-fit tests for the skew-normal distribution when the parameters are estimated from the data. *Commun. Statist. Theory Methods*, **36**, 1735–1755. [87]

Mazzuco, S. and Scarpa, B. 2013. Fitting age-specific fertility rates by a flexible generalized skew-normal probability density function. *J. R. Statist. Soc., ser. A*, under revision. [217]

McLachlan, G. J. and Peel, D. 2000. *Finite Mixture Models*. New York: J. Wiley & Sons. [221]

Meeusen, W. and van den Broeck, J. 1977. Efficiency estimation from Cobb–Douglas production function with composed error. *Int. Econ. Rev.*, **18**, 435–444. [91]

Meintanis, S. G. 2007. A Kolmogorov–Smirnov type test for skew normal distributions based on the empirical moment generating function. *J. Statist. Plann. Inference*, **137**, 2681–2688. 5th St. Petersburg Workshop on Simulation. [87]

Meintanis, S. G. and Hlávka, Z. 2010. Goodness-of-fit tests for bivariate and multivariate skew-normal distributions. *Scand. J. Statist.*, **37**, 701–714. [146]

Meucci, A. 2006. Beyond Black–Litterman: views on non-normal markets. *Risk Magazine*, **19**, 87–92. [186]

Minozzo, M. and Ferracuti, L. 2012. On the existence of some skew-normal stationary processes. *Chil. J. Statist.*, **3**, 159–172. [224]

Montenegro, L. C., Lachos, V. H., and Bolfarine, H. 2009. Local influence analysis for skew-normal linear mixed models. *Commun. Statist. Theory Methods*, **38**, 484–496. [220]

Mudholkar, G. S. and Hutson, A. D. 2000. The epsilon-skew-normal distribution for analysing near-normal data. *J. Statist. Plann. Inference*, **83**, 291–309. [22]

Nagaraja, H. N. 1982. A note on linear functions of ordered correlated normal random variables. *Biometrika*, **69**, 284–285. [52]

Nathoo, F. S. 2010. Space–time regression modeling of tree growth using the skew-*t* distribution. *Environmetrics*, **21**, 817–833. [220]

Naveau, P., Genton, M. G., and Ammann, C. 2004. Time series analysis with a skewed Kalman filter. Chap. 15, pages 259–278 of: Genton, M. G. (ed.), *Skew-elliptical Distributions and their Applications: A Journey Beyond Normality*. Boca Raton, FL: Chapman & Hall/CRC. [224]

Naveau, P., Genton, M. G., and Shen, X. 2005. A skewed Kalman filter. *J. Multiv. Anal.*, **94**, 382–400. [224]

Nelson, L. S. 1964. The sum of values from a normal and a truncated normal distribution. *Technometrics*, **6**, 469–471. [42]

O'Hagan, A. and Leonard, T. 1976. Bayes estimation subject to uncertainty about parameter constraints. *Biometrika*, **63**, 201–202. [42]

Owen, D. B. 1956. Tables for computing bivariate normal probabilities. *Ann. Math. Statist.*, **27**, 1075–1090. [34, 234, 235]

Owen, D. B. 1957. *The bivariate normal probability distribution*. Tech. rept. SC-3831 (TR), Systems Analysis. Sandia Corporation. Available from the Office of Technical Services, Dept. of Commerce, Washington 25, D.C. [235]

Pacillo, S. 2012. Selection of conditional independence graph models when the distribution is extended skew normal. *Chil. J. Statist.*, **3**, 183–194. [158]

Padoan, S. A. 2011. Multivariate extreme models based on underlying skew-*t* and skew-normal distributions. *J. Multiv. Anal.*, **102**, 977–991. [53, 122, 193]

Pérez Rodríguez, P., and Villaseñor Alva, J. A. 2010. On testing the skew normal hypothesis. *J. Statist. Plann. Inference*, **140**, 3148–3159. [87]

Pewsey, A. 2000a. Problems of inference for Azzalini's skew-normal distribution. *J. Appl. Statist.*, **27**, 859–870. [76]

Pewsey, A. 2000b. The wrapped skew-normal distribution on the circle. *Commun. Statist. Theory Methods*, **29**, 2459–2472. [51, 208]

Pewsey, A. 2003. The characteristic functions of the skew-normal and wrapped skew-normal distributions. Pages 4383–4386 of: *XXVII Congreso Nacional de Estadística e Investigación Operativa*. SEIO, Lleida (España). [51, 208]

Pewsey, A. 2006a. Modelling asymmetrically distributed circular data using the wrapped skew-normal distribution. *Environ. Ecol. Statist.*, **13**, 257–269. [208]

Pewsey, A. 2006b. Some observations on a simple means of generating skew distributions. Pages 75–84 of: Balakrishnan, N., Castillo, E., and Sarabia, J. M. (eds), *Advances in Distribution Theory, Order Statistics and Inference*. Boston, MA: Birkhäuser. [94, 188]

Potgieter, C. J. and Genton, M. G. 2013. Characteristic function-based semiparametric inference for skew-symmetric models. *Scand. J. Statist.*, **40**, 471–490. Available online 26 December 2012. [207]

Pourahmadi, M. 2007. Skew-normal ARMA models with nonlinear heteroscedastic predictors. *Commun. Statist. Theory Methods*, **36**, 1803–1819. [222]

Pyne, S., Hu, X., Wang, K., Rossin, E., Lin, T.-I., Maier, L. M., et al. 2009. Automated high-dimensional flow cytometric data analysis. *PNAS*, **106**, 8519–8524. [221]

R Development Core Team. 2011. *R: A Language and Environment for Statistical Computing*. R Foundation for Statistical Computing, Vienna, Austria. ISBN 3-900051-07-0. [75]

Rao, C. R. 1973. *Linear Statistical Inference and its Applications*, 2nd edn. New York: J. Wiley & Sons. [137]

Roberts, C. 1966. A correlation model useful in the study of twins. *J. Amer. Statist. Assoc.*, **61**, 1184–1190. [42, 54]

Rotnitzky, A., Cox, D. R., Bottai, M., and Robins, J. 2000. Likelihood-based inference with singular information matrix. *Bernoulli*, **6**, 243–284. [68, 69, 72]

Sahu, S. K. and Dey, D. K. 2004. On a Bayesian multivariate survival model with a skewed frailty. Chap. 19, pages 321–338 of: Genton, M. G. (ed.), *Skew-elliptical Distributions and their Applications: A Journey Beyond Normality*. Boca Raton, FL: Chapman & Hall/CRC. [192, 228]

Sahu, K., Dey, D. K., and Branco, M. D. 2003. A new class of multivariate skew distributions with applications to Bayesian regression models. *Canad. J. Statist.*, **31**, 129–150. Corrigendum: vol. 37 (2009), 301–302. [190, 192, 194, 200]

Salvan, A. 1986. Test localmente più potenti tra gli invarianti per la verifica dell'ipotesi di normalità. Pages 173–179 of: *Atti della XXXIII Riunione Scientifica della Società Italiana di Statistica*, vol. II. Bari: Cacucci. [86]

Sartori, N. 2006. Bias prevention of maximum likelihood estimates for scalar skew normal and skew *t* distributions. *J. Statist. Plann. Inference*, **136**, 4259–4275. [79]

Serfling, R. 2006. Multivariate symmetry and asymmetry. Pages 5338–5345 of: Kotz, S., Balakrishnan, N., Read, C. B., and Vidakovic, B. (eds), *Encyclopedia of Statistical Sciences*, II edn, vol. 8. New York: J. Wiley & Sons. [2]

Sharafi, M. and Behboodian, J. 2008. The Balakrishnan skew-normal density. *Statist. Papers*, **49**, 769–778. [202]

Shun, Z., Lan, K. K. G., and Soo, Y. 2008. Interim treatment selection using the normal approximation approach in clinical trials. *Statist. Med.*, **27**, 597–618. [225]

Šidák, Z. 1967. Rectangular confidence regions for the means of multivariate normal distributions. *J. Amer. Statist. Assoc.*, **62**, 626–633. [166]

Soriani, N. 2007. *La Distribuzione t Asimmetrica: Analisi Discriminante e Regioni di Tollerenza*. Tesi di laurea, Facoltà di Scienze Statistiche, Università di Padova. http://tesi.cab.unipd.it/7115/. [179]

Stanghellini, E. and Wermuth, N. 2005. On the identification of path analysis models with one hidden variable. *Biometrika*, **92**, 337–350. [158]

Stingo, F. C., Stanghellini, E., and Capobianco, R. 2011. On the estimation of a binary response model in a selected population. *J. Statist. Plann. Inference*, **141**, 3293–3303. [227]

Subbotin, M. T. 1923. On the law of frequency of error. *Mat. Sbornik*, **31**, 296–301. [96]

Tchumtchoua, S. and Dey, D. K. 2007. Bayesian estimation of stochastic frontier models with multivariate skew *t* error terms. *Commun. Statist. Theory Methods*, **36**, 907–916. [192, 225]

Thompson, K. R. and Shen, Y. 2004. Coastal flooding and the multivariate skew-*t* distribution. Chap. 14, pages 243–258 of: Genton, M. G. (ed.), *Skew-elliptical Distributions and their Applications: A Journey Beyond Normality*. Boca Raton, FL: Chapman & Hall/CRC. [186]

Tong, H. 1990. *Non-linear Time Series: A Dynamical System Approach*. Oxford: Oxford University Press. [223]

Tsai, T.-R. 2007. Skew normal distribution and the design of control charts for averages. *Int. J. Rel. Qual. Saf. Eng.*, **14**, 49–63. [225]

Tyler, D. E., Critchley, F., Dümbgen, L., and Oja, H. 2009. Invariant co-ordinate selection (with discussion). *J. R. Statist. Soc., ser. B*, **71**, 549–692. [160]

Umbach, D. 2006. Some moment relationships for skew-symmetric distributions. *Statist. Probab. Lett.*, **76**, 507–512. [11]

Umbach, D. 2007. The effect of the skewing distribution on skew-symmetric families. *Soochow Journal of Mathematics*, **33**, 657–668. [47]

Umbach, D. and Jammalamadaka, S. R. 2009. Building asymmetry into circular distributions. *Statist. Probab. Lett.*, **79**, 659–663. [208, 210]

Umbach, D. and Jammalamadaka, S. R. 2010. Some moment properties of skew-symmetric circular distributions. *Metron*, **LXVIII**, 265–273. [208]

Van Oost, K., Van Muysen, W., Govers, G., Heckrath, G., Quine, T. A., and Poesen, J. 2003. Simulation of the redistribution of soil by tillage on complex topographies. *European J. Soil Sci.*, **54**, 63–76. [160]

Vernic, R. 2006. Multivariate skew-normal distributions with applications in insurance. *Insurance: Math. Econ.*, **38**, 413–426. [159, 160]

Vianelli, S. 1963. La misura della variabilità condizionata in uno schema generale delle curve normali di frequenza. *Statistica*, **33**, 447–474. [96]

Walls, W. D. 2005. Modeling heavy tails and skewness in film returns. *Appl. Financial Econ.*, **15**, 1181–1188. [119]

Wang, J. and Genton, M. G. 2006. The multivariate skew-slash distribution. *J. Statist. Plann. Inference*, **136**, 209–220. [195]

Wang, J., Boyer, J., and Genton, M. G. 2004. A skew-symmetric representation of multivariate distributions. *Statist. Sinica*, **14**, 1259–1270. [11, 175]

Weinstein, M. A. 1964. The sum of values from a normal and a truncated normal distribution. *Technometrics*, **6**, 104–105. [42]

Whitt, W. 2006. Stochastic ordering. Pages 8260–8264 of: Kotz, S., Balakrishnan, N., Read, C. B., and Vidakovic, B. (eds), *Encyclopedia of Statistical Sciences*, II edn, vol. 13. New York: J. Wiley & Sons. [10]

Yohai, V. J. 1987. High breakdown-point and high efficiency robust estimates for regression. *Ann. Statist.*, **15**, 642–656. [112, 114, 115]

Zacks, S. 1981. *Parametric Statistical Inference*. Oxford: Pergamon Press. [26]

Zhang, H. and El-Shaarawi, A. 2010. On spatial skew-Gaussian processes and applications. *Environmetrics*, **21**, 33–47. Available online 17 March 2009. [223]

Zhou, T. and He, X. 2008. Three-step estimation in linear mixed models with skew-*t* distributions. *J. Statist. Plann. Inference*, **138**, 1542–1555. [220]

Index

256

Printed in the United States
By Bookmasters